Surface Pressure Prediction in High-Speed Flows

Thimmer

TABLE OF CONTENTS

CHAPTER 1

INTRODUCTION

Studies of supersonic and hypersonic flow regimes formulate important aspects of the science of space. The research is concerned primarily with space science, advanced weaponry, and spacecraft. The defense industry utilizes the supersonic projectiles- the rockets, missiles, and aircraft. Supersonic wings are usually designed in wedge shape or in delta wings for use in the supersonic vehicle category. A supersonic projectile is one in which the flow behavior over wedges and delta wings are researched at high angles of incidence. Present research is directed toward improving the understanding of supersonic/hypersonic flow fields around cones, wedges, ogives, and flat deltas in a very broad range of angles of attack for the respective development of space shuttles and advanced military aircraft. A clear and accurate method will be required in parametric calculation towards understanding aerodynamic loading and stability for these types to meet the process of design. Apart from this, high-angle-of-attack dynamic stability computation for these air forms is still a challenge at present. In a descent mode, shockwaves usually become very strong and might attach to or detach from the vehicle.

There is a pronounced need for precision analytical solutions nowadays, applied in aerospace high-tech situations. The astronomical cost of experimental facilities can hardly justify their establishment unless analytical methods are developed to the desired level of accuracy. This is the kind of ability that allows consideration of the delta-wing aerodynamic characteristics for a broad range of Mach numbers. We applied this analytical method in analyzing the stability of aerospace vehicles to assess or compare the accuracy of our results. To design effective hypersonic/supersonic vehicles, it is imperative to fully comprehend the interaction between the aerodynamic performance and the geometrical configuration parameters. In order to make accurate parametric calculations, designers must understand the aerodynamic load and stability aspects for these configurations. The primary aspect we deal with concerns dynamic stability evaluations regarding these configurations during re-entry trajectories or while maneuvering at high angles of attack.

The linearized theory applies to supersonic flow at small angles of incidence. At transonic and hypersonic speeds, the linearization is said to fail. For transonic flow, as the free stream Mach number approaches unity, the perturbation velocity plus its first derivative in the free stream direction cannot be neglected, and thus the budding violates the linearized assumption. In hypersonic flow, perturbation velocities are to a greater extent smaller than the free stream velocities. Even though the pressure variations are to be much lesser than the dynamic pressure of the free stream, when compared to the static pressure of free stream, these cannot be regarded as negligible quantities. So, the disturbances in the true sense cannot be called small-the way linearized theory has held in the past. The linearization will thus be valid under the other assumption that the disturbance generated by the body plays a weak role. Hence, these linearizations approximate waves akin to the Mach lines that are in a supersonic set-up. The concept of the silent zone describes that hypersonic flow is continuously passed ahead of the nose Mach cone in which disturbances have yet to exist. The Mach angle at extremely high Mach numbers tends to coincide with the maximum deflection angle for slender bodies and aerofoils. It is also accepted that disturbances and their gradients essentially work out at right angles to the free stream. This feature is the basis of small perturbation theory and also establishes its similarity. In contrast, hypersonic flows prove an exception, which condition has not at all been addressed. Relevant equations of motion and boundary conditions will need to be framed for further solution.

Ghosh and Mistry [25] propounded a theory as seen with quasi-steady flow over an oscillating two-dimensional wedge, which later, Ghosh broadened with regard to the axi-symmetric flow issues Ghosh [26]. Two-dimensional oscillating wedge problems were studied by Appleton, who also considered the effect of secondary waves. This current undertaking takes the form of an analytical and computational analysis of pressure at the apex of a 2D planar wedge at high Mach numbers and with different angles of incidence. We have, in addition, incorporated studies related to the computational pressure distribution on the surface of the 2D wedge under supersonic and hypersonic flows of reasonably high intensity. Besides, some aspects of Ghosh's [24] large deflection similitude, which has been utilized for the construction of an unsteady piston theory theorizing the stability derivatives for oscillating planes and wedges were also investigated. Adapting Ghosh's [26] large deflection hypersonic similitude, the current study predicts the stability derivatives in pitch for supersonic and hypersonic flows using a wider range of Mach numbers, incidence angles, and pivot points than previously.

Further this study also extended to computational investigation of the dimensionless surface pressure along the edge of 2D planar wedge at supersonic and hypersonic Mach numbers for considerable range of angle of incidence. Some of these investigations also try to look in detail on the numerical simulation of the coefficients associated with stiffness and damping derived across the wedge at varying angle but at the same time considering different pivot positions in both supersonic and hypersonic Mach numbers. Estimation of the derivatives was done using regression model approach of analysis in respect to stiffness and damping derivatives. The present work also mainly aims at a computational study of the surface pressure and thermal development along the slant length of the 3D cone under both the supersonic and hypersonic flow realm using the Computational Fluid Dynamics modeling with other numerical analysis set using regression processing modeling.

CHAPTER 2

LITERATURE REVIEW

Tsien [1]is credited with the similarity rule for unstable and irrotational hypersonic flow. Results from his studies correlate very well with experimental results. Hayes [2] dealt with the case of unstable high Mach flows over a very thin aerofoil. In this research, similar to the previous work of Tsien [1]the focus is limited to irrotational movements and will encounter the same error as Tsien [1] concerning the mean-surface approximation of a boundary condition. In the steady-state flows, Hayes [2] indicated that equivalence principle is suitable. Morgan H G et.al [3] formulated one of the first comprehensive small disturbance aerodynamic models able to predict aeroelastic flutter in the transonic and supersonic regimes. Their work was able to extend the classical linear theory by introducing Mach-dependent pressure distributions and shock-induced changes in aerodynamic stiffness and damping. Hayes and Probstein [4] emphasized that Hayes's equivalence principle can be modified for unsteady flow. Zartarian et al.[5] applied tanged-wedge hypothesis and shock wave expansion concept to study unsteady hypersonic flow. These studies set the foundation for investigating flow over thin bodies, as studied by Sychev [6], who offered perspectives on 3-D hypersonic gas flows around slender bodies at high angles of attack, a difficult situation in hypersonic research. Cole and Brainerd [7] attempted to fill gaps in understanding slender, delta-type wings at high angles of attack in hypersonic flow. Pressure characteristics on delta wings with shock attachment were studied by Pike [8], who made suitable correlating work with that of Hui [9] on stability in oscillating wedge and caret wing models at high speeds

Hui's[10] research is contributing immensely in development of unified theory in supersonic and hypersonic flow over the lower surface of delta wing in attached shockwave case. Just like Orlik-Ruckemann's study [11] on achieving the stability requirements of the aircraft. Carrier [12] and Hui[13] dealt with exact solution of 2D flow for an oscillating wedge and flat plat respectively in the view of attached shock wave at supersonic speed and angle of incidence. The attached or detached orientation of the shock wave with respect to the leading edges for an oscillating triangular wing would mainly depend on the flying of Mach number, angle of attack, gas specific heats, and the backward swept angle of the wing, whether it is supersonic or hypersonic. The attached-shock case of oscillating delta wings was examined by Liu and Hui[14] whereas Hui and Hemdan [15] considered the unsteady flow with a detached shock wave in the view of thin shock layer theory. In the analysis of aerofoils undergoing large flow deflection, Ericsson's [16] takes into account both viscous and elastic forces. At zero angle of attack Rossow [17] introduced a rational parameter to evaluate the pressure distribution on 3D bodied in hypersonic flow.

2.1 High speed flows aerodynamics

The unsteady two-dimensional flow problem may be correlated to the motion of gas in a tube pushed by a piston- this was introduced by Light Hill [18]. He obtained the aerodynamic derivatives of a symmetric pitching aerofoil using analytical expression based on a perfect gas assumption. The unsteady shock expansion theory by Hayes and Probstein [4] and the widely-accepted piston theory of Lighthill[18] no consideration is given to the effects of any secondary waves arising from the surface of the body and reflected from the bow shock. A key requirement in the undisturbed fluid is that the piston's velocity stays below the speed of sound. In unsteady flow, Ashley et al., [19] estimated aerodynamic forces by extending Lighthill's [18] piston theory. The piston theory developed by Lighthill [18] has no reference to the secondary wave reflection effects. These effects have been included by Appleton [20] and McInthosh [21] who independently demonstrate that strong shock piston theory is reduced when neglecting these effects. Landahl [22] treated an unsteady aerodynamic response of a thin-wing undergoing harmonic oscillations in a supersonic flow. Using piston theory for small-disturbance potential flow, he derived analytical expressions for the pressure field induced by gusts and wing motions. The piston theory is then incorporated into Mile's [23] shock-expansion theory by Ghosh [24] to obtain a second-order approximation (order $Ø^2$) with respect to Ø, which is the angle between a plane approximating the windward surface and an attached shock. Later, Ghosh and Mistry [25] advanced this analogy for oscillating delta wings at hypersonic Mach numbers and high angles of incidence, in the presence of attached shocks. In addition, Ghosh [26] investigated similitude principles for large deflections in quasi-cones and quasi-wedges. By investigating the effects of massive deflections on these shapes in hypersonic flows, this study expanded on earlier research and allowed for a broader use of similitude approaches. Hemdan [27] studied the unsteady aerodynamic response of two-dimensional hypersonic airfoils subjected to small amplitude oscillations at low angles of attack. The derived expressions for unsteady lift and moment coefficients using linearized potential flow theory and piston theory approximations. With the help of computational fluid dynamics (CFD) simulations and an analytic model, Tarpley & Lewis [28] determine static and dynamic stability derivatives at high

Mach number conditions. The results revealed that the geometric design has a counterpart on aerodynamic stability.

2.2 Stability Derivatives in Supersonic/Hypersonic Regimes

A unified panel method has been developed to study both hypersonic and supersonic aeroelastic behavior for arbitrary geometry by Chen and Liu [29]. In the case of blunt wedges, Balakumar [30] studies the linear stability. It highlights the influence of wedge bluntness and Mach number on the types of instability modes, providing insight into the mechanisms involved in transitioning in high-speed flows. In a follow-up investigation, Balakumar and Owens [31] analyzed the stability of hypersonic boundary layers on a sharp cone. The study conducted by Khan and Crasta [32] examined the aerodynamic behavior of a delta wing. The investigation indicated that the leading-edge curvature had an influence upon the unsteady aerodynamic characteristics. Anton et al. [33] established a benchmark in the validation of stability derivatives against the X-31 aircraft through combined wind tunnel tests and computational methods for delta-wing configurations. Da Ronch et al. [34] went on to employ CFD as "a tool to investigate dynamic stability derivatives" and showed that numerical methods could be viable replacements for wind tunnel data in some regimes. Rom [35] had an excellent text on generally high angle-of-attack aerodynamics over speed regimes with detailed considerations of flow separation and control. Asha Crasta and Khan S A [36] have studied the behavior of supersonic flows over a planar wedge for high incidence angles. Firouz-Abadi & Alavi [37] discussed fin thickness and angle-of-attack effects on aeroelastic stability at supersonic flows with particular focus on sensitivity of designs. Kazemba et. al [38] provided a survey about dynamic stability of blunt bodies in supersonic flow, identifying the main physical mechanisms theoretically involved in the process. Mader & Martins [39] presented a novel method for determining performance derivatives gradients for optimization purposes, nonetheless proved helpful for design loops. The similitude concept to hypersonic flows over planar wedges was extended by A. Crasta and S A Khan [40] in their subsequent study. In a later study, Asha C and Khan S A [41-42] investigated the influence of angle of incidence on the stability derivatives of delta wings with curved and straight leading edges. Crasta & Khan [41] focused on delta wings under Newtonian approximation in terms of pressure distribution and stability in flow speeds. Crasta & Khan[42] investigated superimposed supersonic similitude for non-planar wedge geometries, concerning oscillatory motion. Crasta et al. [43] studied the influence of Mach number and angle of attack on the variation of stiffness derivatives, representing dynamic response trends. Asha Crasta et al. [44] also examined the surface pressure distribution. Shabana et al. [45] have investigated pressure and its derivatives for oscillating cones considered in dynamic response modeling. Monis et al. [46] had done the evaluation of damping derivative under supersonic conditions, which is essential for flight control and stability checks. Barua et al. [47] carried out numerical simulations to investigate shock oscillations over a biconvex airfoil and found that there existed local geometric perturbation-related effects. Musavir et al., [48] studied the investigation of aerodynamic derivatives by employing computational and analytical method for oscillating delta wing model at hypersonic Mach number. Besides these, with $\gamma = 1.66$ Shabana et al. [49] have derived the stability derivatives of oscillating cones, to emphasize more the effects of gas dynamics at high-speed flows. Shabana et al. [50] included semi-vertex angle effects on stability derivatives, simply because this will assist in conical body design. Khan et al.[51] studied the numerous flow variable in the case of wedge. Monis et al.[52] evaluated stiffness derivatives for delta wings. Mi & Zhan [53] presented a critical review of numerical approaches for dynamic stability derivatives and advocated hybrid CFD and reduced-order modeling strategies. Monis et al. [54] carried out an analytical evaluation of stability derivatives. Aysha Shabana et al. [55] investigated the stiffness derivatives for ogive-shaped bodies at a constant specific heat ratio 1.666. The damping derivatives for ogives under high-speed flow conditions were investigated by Aysha Shabana et al. [56] in a follow-up study. In fact, Renita Monis et. al. [57] were investigating the difference between quasi-steady and unsteady damping derivatives for delta wings operating in hypersonic conditions. Shamitha et al. [58] studied the pressure over an oscillating wedge at hypersonic speeds by combining analytical and CFD method and optimizing by Taguchi's method. An experimental study was done by Baidya et al. [59] pertaining to cylinder wakes from subsonic to supersonic regimes, where the transition behavior in wake structures was identified. Shamitha et al. [60] explores the aerodynamic damping characteristics of non-planar wedge configurations in the context of Newtonian flow condition. In the further study, Shamitha et al. [61] investigate how varying the sweptback angle of delta wings affects the pressure distribution across their surfaces at supersonic speeds.

2.3 Suddenly Expanded Flows and Their Characteristics

Rathakrishnan et al. [62] explored the introduction of cavities in the base region changed the flow field in sudden-expiration configurations. The presence of cavities changed the base pressure and its shock structure, thereby stabilizing the flow. Rathakrishnan [63] studied the effect of ribs on base pressure. The ribs generated stream wise vortices created by their presence in the flow, enhance mixing, change pressure distribution, and prevent flow separation. Rathakrishnan & Khan [64] were the first to use active control with micro-jets for orbital nozzles. They found that injecting jets into the neighborhood of the base caused an increase in base pressure and attenuation of shock oscillations. Microjets also improve base pressure regulation at low momentum inputs, and are thus most effective in an over expanded regime, as shown by Khan and Rathakrishnan [65]. Khan & Rathakrishnan [66] addressed situations of under expanded flow in which the control effectiveness is shown to be dependent solely on the nozzle pressure ratio (NPR) and the jet location. Khan & Rathakrishnan [67] described the effects of correctly expanded flows in which microjets improved symmetry and base pressure even without shock separation. In Part II, Khan & Rathakrishnan [68] further elaborated on the parametric influences such as the jet diameter and the pressure ratio and location regarding control of under expanded flow. Khan & Rathakrishnan [69] considered the levels of nozzle expansion systematically reiterating that for maximum control effectiveness, optimal NPR and jet configuration are imperative. In further study, Khan S A and Rathakrishnan [70] have studied the active control of base pressure in supersonic conditions and provided fundamental insight into the subject. The behavior of the jet streams to actively control low-Pressure recirculation region just downstream of the sudden expansion has been investigated by Rehman & Khan [71]. Kalimuthu et al. [72] have used spiked bodies in Mach-6 hypersonic flow, which were shown to reduce fore body pressure and drag experimentally. Kalimuthu et al. [73] further quantified drag reduction mechanisms on spiked blunt-nosed bodies and verified substantial benefits for Mach 6. M Ahmed Ali Baig et al. [74] carried out experimental studies at an area ratio of 6.25 for over-expanded flows, thus confirming the sensitivity of base pressure to geometry and NPR. Khan et.al [75] revealed base flow behavior in a suddenly expanded nozzle by wall pressure measurements. Ullah M A et.al [76] examined the active control of wall pressure fields at low supersonic flows. Using controlled microjets, they have shown that secondary flows introduced at a strategic location can reduce shock-induced disturbances and suppress flow separation. Khan, [77] examined base pressure control. In supersonic flow regime, Quadros et.al [78] conducted a numerical investigation of suddenly expanded flows. Khan et al. [79] analyzed the converging-diverging (CD) nozzles using CFD Saleel et al. [80] have described an experimental investigation based on the base pressure characteristics of a sudden expansion nozzle. To reduce the drag, Khan et al. [81] studied the CFD analysis of human powered submarine. Experimental examinations of nozzle flow experiencing sudden expansions with Mach 1.1 were conducted by Faheem et al. [82]. Their study focused on the behavior of under-expanded super-sonic jets and the influences of area ratios and nozzle pressure ratios on the shock structures and flow separations. Kalimuthu et al. [83] studied the aerodynamic behavior of blunt spiked bodies. Zuhair and Mohammed [84] experimented with various trailing edge geometries and their effect on the aerodynamic performance of low-speed Blended Wing Body (BWB) aerial vehicles. They employed computational and experimental techniques to analyze the performance of a number of trailing edge geometries, demonstrating that the geometry greatly influenced the lift-to-drag ratio and stability. Azami et al.[85] measured wall pressure distributions and jetting effects at Mach 1.5, demonstrating how active flow control concepts can affect aerodynamic stability and efficiency. Pathan et al. [86-89], discussed changes in base pressure in rapidly expanded flows with Computational Fluid Dynamics (CFD) and control jet techniques. Their investigations have demonstrated how the placement of control jets and the pressure ratios used for blowing have affected base pressure and flow reattachment. An earlier study by Khan and Rathakrishnan [70] attempted active manipulation of base pressure in supersonic conditions with a view toward establishing a basic understanding in this area. Further experimenting based on these principles by Pathan et al., [88-89], [90], [91,93] have attempted investigations into nozzle pressure ratios, expansion levels, and duct optimization. These attempts increased the understanding of thrust dynamics and base pressure control in fast-moving flows. Sethuraman et al. [92] basically gave an overview of pressure control methods involving cavities and ribs, summarizing their effectiveness as passive devices in altering wall and base pressures. Pathan et al., [93] conducted a comprehensive investigation into the aerodynamic performance of a boat-tail helmet design aimed at reducing drag. Shaikh et al., [94] aimed to achieve uniform flow and reduce pressure drop across the monolith. The implications of their research highlight the significance of uniform flow distribution. Khan et al. [95] conducted an extensive review of passive control mechanisms for decreasing base pressure in high-velocity flows. Meng et al. [96] experiment on combined spike with multi-jet configuration with double cone missile for drag reduction and thermal protective investigation. Shaikh and Havaldar [97] conducted a comprehensive review of the methods for controlling the base pressure in fluid dynamics, specifically on active and passive control techniques. Aqilah et al., [98] also support the effectiveness of passive control mechanisms. They also suggest that further investigation is needed into the complexities of high-speed aerodynamics. In sonic speed, Khan et al.[99] study and examined the effect of ribs. Pathan et al., [100] emphasize the various benefits of geometric optimization in aerospace engineering. Nurhanis et al. [101] extended active control into the supersonic regime. This combination of micro-jet actuation and span wise distributed ports.

Fakhruddin et al., [102] investigated and analyzed the base pressure control with ribs at Mach number M=1.2 using CFD. Khan et al., [103] studied the effect of rib size, shape, and location in a suddenly expanded flow at sonic Mach number. Subramani et. al [104] performed a numerical study on the passive types of geometries such as tabs, chevrons, and lobed nozzles that affect jet mixing at supersonic velocities. Ismail et al. [105] carried out an experiment on the aerodynamic effect of cavities in sudden expansion into the supersonic flow, showing that cavity geometry considerably influences base pressure and flow structure. Ridwan et al. [106] continued to investigate how cavities might be effectively employed to control base pressure in flows from a 5.29 area-ratio converging-diverging nozzle with good pressure recovery. A comprehensive CFD analysis by Shetty et al. [107] at Mach 1.3 showed that placing a quarter rib could serve as an effective passive control device in stabilizing base pressure and reducing flow separation.

2.4 Research Gap

From the literature it is found that Crasta et. al [44] estimated the surface pressure distribution of 2D wedge using only by analytical method. In the present study, surface pressure distribution at the nose of two dimensional wedge investigated by applying both analytical and computational techniques. For the computational investigation of surface pressure distribution at the nose of 2D wedge, computational fluid dynamics (CFD) analysis is adopted and then results were compared with analytical findings.

Validation of stiffness and damping coefficient analytical results by computational methods is the subject of the current investigation. The stiffness and damping coefficients are simulated numerically during regression model analysis. Full factorial design of experiment methodology is considered in numerical simulation to get stiffness and damping coefficient. These are some more advantages of this research work as compared to previously described studies in the literature.

It was also found from literature that many investigations contributed towards high-speed aerodynamics flows, but there were not much studies available on surface pressure distribution over the edges of a two-dimensional wedge in high-speed flows. In this study, the surface pressure distribution along the edge of steep two-dimensional wedges is explored through computational fluid dynamics (CFD) analysis for a significant range of Mach numbers, angles of incidence and x/L ratios. In the present study, a detailed regression analysis is conducted for the parameters Mach number, angle of incidence and pivot position and concluded with the remarkable conclusions based on the results obtained by analytical and computational methods.

CHAPTER 5

AERODYNAMIC DERIVATIVES FOR AN OSCILLATING WEDGE

Studies of supersonic and hypersonic flow regimes formulate important aspects of the science of space. The research is concerned primarily with space science, advanced weaponry, and spacecraft. The defense industry utilizes the supersonic projectiles- the rockets, missiles, and aircraft. Supersonic wings are usually designed in wedge shape or in delta wings for use in the supersonic vehicle category. The current study, investigates look in detail on the numerical simulation of the coefficients associated with stiffness and damping derived across the wedge at varying angle but at the same time considering different pivot positions in both supersonic and hypersonic Mach numbers. Estimation of the derivatives was done using regression model approach of analysis in respect to stiffness and damping derivatives.

Table 5. 1 Parameters considered for the aerodynamic derivatives for an oscillating wedge

Parameter	Levels	Values
Mach Number (M)	21	2.2, 2.4, 2.6, 2.8, 3.0, 3.2, 3.4, 3.6, 3.8, 4.0, 5, 5.5, 6.0, 6.5, 7.0, 7.5, 8.0, 8.5, 9.0, 9.5, 10.0
Angle (θ)	18	2°, 3°, 4°, 5°, 6°,8°, 9°, 10°, 12°, 14°, 15°, 16°, 18°, 20°, 21°, 24°, 27° and 30°
Pivot Position (h)	06	0.0, 0.2, 0.4, 0.6, 0.8, and 1.0.

5.1 Evaluation of Stiffness Derivative of a 2D Wedge at Supersonic Speeds

Analysis:

Fig. 5.1.1 illustrates the geometry of the planar wedge transfer of the pivot point.

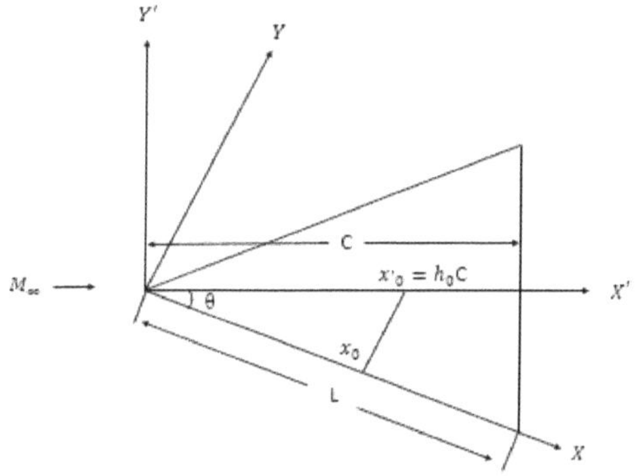

Fig. 5.1. 1 Two-dimensional geometry of a wedge

The low amplitude oscillations around the pivot position in pitch should be seen by the flat plate aerofoil, specified by its length (L) and wedge angle (θ). The variable x_0 indicates the distance from the apex. Eq. (5.1.1) provides the piston velocity at any given moment, considering the distance x and the angle of attack α.

$$U_p = U_\infty \sin\alpha + q(x - x_0) \tag{5.1.1}$$

The expression given by Eq. (5.1.2) indicates the Mach number 'Mp' of the piston. Thus,

$$M_p = M_\infty \sin\alpha + \frac{q(x - x_0)}{a_\infty} \tag{5.1.2}$$

Lighthill [18] studied the exact isentropic expression for piston pressure, which is given in a power series correlating with velocity. To maintain the isentropic state, the piston's velocity must be less than or equal to the sound velocity in a free stream. This notion aligns with the principle of minor disturbances that drives Lighthill [18] piston theory.

5.1.1 Piston Theory

In the view of Ghosh [26], the z-direction, where the velocity component is at its lowest, can be considered independent of the wedge strip parallel to the centerline. The combination of strip theory and Ghosh's [26] similitude approach led to the development of the "Piston analogy," tied explicitly to the Mach number "Mp." In this situation, the angle of incidence and piston Mach number are considered satisfactory. Consequently, Ghosh's piston theory is applied instead of Lighthill's [18] or Mile's [23] strong shock expansion hypothesis. The inertia level at the wing's surface piston is directly impacted by the surface pressure (P), as shown in Eq. (5.1.3).

$$\frac{P}{P_\infty} = 1 + A(M_P)^2 + A M_P \sqrt{B + (M_P)^2} \qquad (5.1.3)$$

where $P\infty$ is the free stream pressure.

$A = \frac{\gamma(\gamma+1)}{4}$, $B = \left(\frac{4}{\gamma+1}\right)^2$ and γ is the specific heat ratio of gas.

At different span positions, strips are considered distinct, and the wedge angle coincides with the wing. The strips are individually processed at various locations along the span, with the wedge angle indicating the wing. Both the piston Mach number and flow deviation are allowed to rise now. The incidence angle and the wing angle are equal. The angle 'ø' represents the angle between the shock and the strip, with uniform incidence angle and wing angles. Eq. (5.1.4) displays that the piston analogy has been enlarged to supersonic flow past, and the corresponding expression is provided below.

$$\frac{P}{P_\infty} = 1 + A\left(\frac{M_P}{\cos\phi}\right)^2 + A\left(\frac{M_P}{\cos\phi}\right)\sqrt{1 + B\left(\frac{M_P}{\cos\phi}\right)^2} \qquad (5.1.4)$$

5.1.2 Pitching Moment Derivative

Let the average incidence be referred to as α_0 for a wing oscillating in pitch with low frequency and low amplitude along an axis x_0. The pressure on the windward surface and the piston velocity remain constant over a span-wise strip of length 2L at point x. The nose-down moment is given by Eq. (5.1.5), with the assumption of zero pressure on the lee surface.

$$m = -\int_0^L P(x - x_0)\,dx \qquad (5.1.5)$$

Now the aerodynamic Stiffness derivative is given by,

$$-Cm_{\alpha} = \frac{1}{\frac{1}{2}\rho_{\infty}U_{\infty}^{2}L^{2}}\left(\frac{-\partial m}{\partial \alpha}\right)_{\alpha=\alpha_0, q=0} \tag{5.1.6}$$

By equating Eq. (5.1.2) and Eq. (5.1.4) in Eq. (5.1.5) and differentiation under integral sign is applied. By defining, $x_0 = hL\cos^2\alpha_0, C = L\cos\alpha_0$ and $S_1^1 = \frac{M_{\infty}\sin\alpha_0}{\cos\phi}$ the required Stiffness derivatives in pitch of a planar wedge will be of the form,

$$-Cm_{\alpha} = \left[\frac{\gamma+1}{M_{\infty}\cos\alpha_0\cos\phi}\right] F(S_1^1)\left[\frac{1}{2} - h\cos^2\alpha_0\right] \tag{5.1.7}$$

Where,

$$F(S_1^1) = 2S_1^1 + \frac{B + 2(S_1^1)^2}{\sqrt{B + (S_1^1)^2}} \tag{5.1.8}$$

5.1.3 Design of Experiment

Design of experiment (DOE) is a scientific and systematic strategy that allows you to investigate the link between various input variables and the intended outcome. The Design of the Experiment (DOE) is a powerful data collection and analysis tool in different experimental situations. It allows the adjustment of several input factors to determine the impact on the desired result. DOE can identify essential interactions that might be missed when looking at one element at a time by manipulating many inputs simultaneously. DOE covers manufacturing, process improvement, new product development, and engineering decision-making procedures. Businesses in various sectors have embraced DOE, including marketing, chromatography, energy, hospitals, pharmaceuticals, and the food industry. It is feasible to look into every combination that could occur, even only some of them. This methodology applies not only to physical processes but also to computer simulation models. Ultimately, DOE is a versatile tool for identifying crucial input factors and understanding their relationships with output variables across various circumstances. As a scientific approach, DOE has become more prevalent in fields such as engineering, science, technology, medicine, computer science, and accounting, accounting for approximately half of all applications compared to other scientific disciplines.
Specific general procedures involve clearly articulating the objectives and outlining the problems or questions to investigate:

i. Find the response variable that reflects the measurable outcome of the experiment compared with the desired objectives.

ii. Consider the factors and their levels by choosing independent variables that result in a change in the response variable.

iii. Choose the appropriate type of experimental design according to the objectives.

iv. Use the specified design matrix to experiment, ensuring it follows the experimental plan.

v. Analyse the data using statistical approaches like regression and ANOVA to gain relevant insights and find significant components.

vi. Draw practical conclusions and recommendations based on the outcomes of the analysis. Provide graphical representations to help readers visualize the data and validate the outcomes.

vii. Address challenges associated with DOE planning, such as selecting suitable factors for testing (independent variables), determining appropriate ranges for the factors, calculating the required number of replicates, and determining the need for a center point.

The following statistical software packages are available to facilitate the design and analysis of DOE: Minitab, Statistica, SPSS, SAS, Design-Expert, Stat graphics, Prisma, R, and Action for Microsoft Excel. The whole factorial design approach is considered for data analysis in the current investigation. DOE utilizes Mach Number(M), Wedge angle (θ), and Pivot Position (H) as its factors. Designing the experiment involves considering the six levels of pivot position (H) within the combination of Mach number (M) and incidence angle (θ) at level 10. Table 5.2 illustrates the factor, factor level, and the factor values of the data.

Table 5. 2 Full factorial: Level and the factors of the design of experiments

Factor	Type	Levels	Values
Mach Number (M)	Fixed	10	2.2, 2.4, 2.6, 2.8, 3.0, 3.2, 3.4, 3.6, 3.8, 4.0
Wedge Angle (T)	Fixed	10	2, 4, 6, 8, 10, 12, 14, 16, 18, 20
Pivot Position (H)	Fixed	6	0.0, 0.2, 0.4, 0.6, 0.8, 1.0

5.1.4 Results and Discussion

5.1.4.1 Major impact plots of Mach numbers for the Stiffness derivative ($-C_{m_\alpha}$)

Fig. 5.1.2 displays the main impact of the stiffness derivative over the wedge. Figure 2 demonstrates the mean stiffness derivative values and Mach numbers 2.2, 2.4, 2.6, 2.8, 3.0, 3.2, 3.4, 3.6, 3.8, and 4.0 with a varied angle of incidence (θ) ranging from 2° to 20° at all pivot locations h= 0.0 to 1.0. The results imply that there is a drop in the values of the stiffness derivative with increasing Mach number. Additionally, a slight fluctuation in the mean values of stiffness derivatives is noted for Mach numbers 3.0 to 4.0 compared to Mach numbers 2.2 and 2.8.

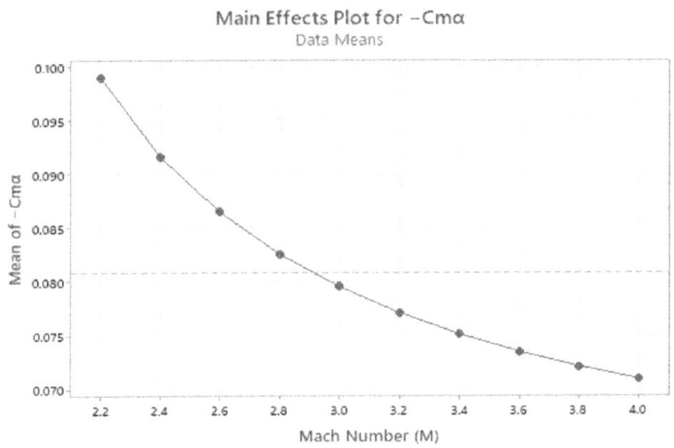

Fig. 5.1. 2 Major impact of Mach numbers on the Stiffness Derivative across the 2D wedge

5.1.4.2 Major impact plots of the angle of incidence for the Stiffness derivative ($-C_{m_\alpha}$)

Fig. 5.1.3 displays the main impact of the stiffness derivative over the wedge. Fig. 5.1.3 illustrates the plot of mean values of stiffness derivative and angle of incidence (θ) ranging from 2° to 20°, with Mach numbers (M) of 2.2, 2.4, 2.6, 2.8, 3.0, 3.2, 3.4, 3.6, 3.8, and 4.0, and pivot positions h from 0.0 to 1.0. According to the findings, the stiffness derivative increases with an increasing angle of incidence. There is a more significant difference in the stiffness derivative for angles of incidence between 10° and 20° compared to angles between 2° and 8°. Consequently, the stiffness derivative shows a gradual rise in value with increasing incidence angles. Consequently, the stiffness derivative's value gradually rises as the incidence angle grows.

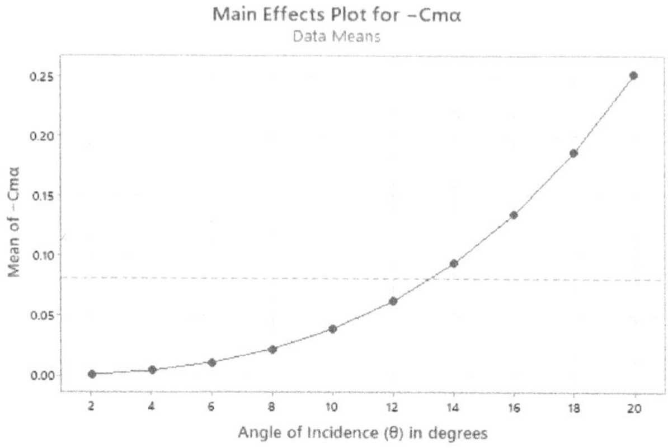

Fig. 5.1. 3 Major Impact of Incidence Angle on Stiffness Derivative across the 2D wedge

5.1.4.3 Major impact plots of pivot position for the Stiffness derivative ($-C_{m_\alpha}$)

Fig. 5.1.4 shows the significant impact of the stiffness derivative on the wedge. Fig. 5.1.4 displays the graph showing the average values of stiffness derivative at pivot points located at 0.0, 0.2, 0.4, 0.6, 0.8, and 1.0. with the variable values of Mach number (M) 2.2, 2.4, 2.6, 2.8, 3.0, 3.2, 3.4, 3.6, 3.8, and 4.0 and the angle of incidence (θ) 2°, 4°, 6°, 8°, 10°, 12°, 14°, 16°, 18°, and 20°. The findings demonstrate a constant drop in the stiffness derivative from the pivot point, h = 0.0, to 1.0. From the pivot position h= 0.0 to 1.0, it is also evident that the stiffness derivative reduction is linear.

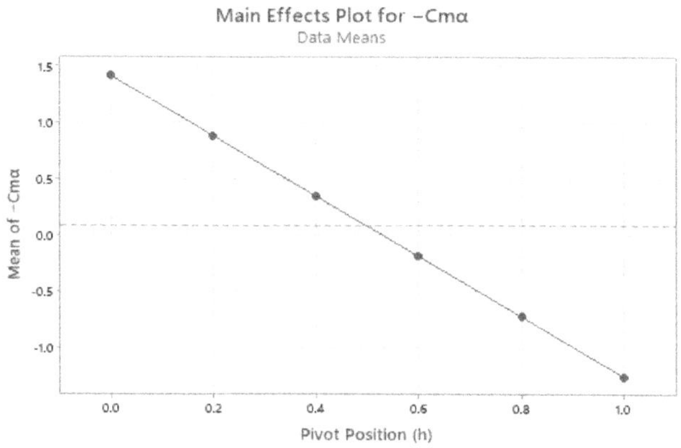

Fig. 5.1. 4 Main effect of Pivot Position on Stiffness Derivative over the 2D wedge

5.1.4.4 Graph of Interaction among Incidence Angle (θ), Mach Number (M), and Pivot Position (h) for Stiffness derivative

Fig. 5.1.5 represents an interaction plot for Mach number (M), incidence angle (θ), and pivot position (h), showing the relationship for stiffness derivative. In the figure, it is observed that the center of pressure differs from h = 0.5 to 0.58, while for the Mach numbers from 2.2 to 4.0, the stiffness derivative decreases from h = 0 to h = 1. The shift of the center of pressure is usually created by the changes in pressure distribution. In this case, there is no further movement of the center of pressure at high Mach numbers. The greater the angle of incidence (θ), the larger is the stiffness derivative absolute value from h = 0 to h = 1. Also, the great semi-vertex angle would push the center of pressure toward the trailing edge.

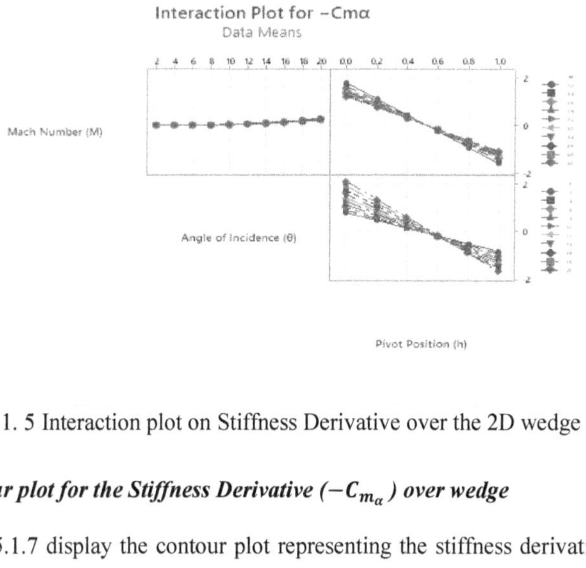

Fig. 5.1. 5 Interaction plot on Stiffness Derivative over the 2D wedge

5.1.4.5 The contour plot for the Stiffness Derivative ($-C_{m_\alpha}$) over wedge

Fig. 5.1.6 and 5.1.7 display the contour plot representing the stiffness derivative across a wedge. Fig. 5.1.6 displays the contour plot of the stiffness derivative concerning Mach Number (M) and Pivot position (h). Fig. 5.1.6 illustrates how the stiffness derivative across a wedge declines as the Mach number rises at every pivot point. Additionally, it is noted that the stiffness derivative becomes lower for each Mach number at the pivot position h = 0.0 to 1.0. Fig. 5.1.7 displays the contour plot for the stiffness derivative with the pivot position (h) and angle of incidence (θ). The Fig. 5.1.7 contour plot shows that higher angles of incidence (θ) lead to more significant stiffness derivatives along the wedge at every pivot point. Additionally, it is observed that the derivative of stiffness increases for every angle of incidence (θ) as the pivot position h ranges from 0.0 to 1.0. The Appendix – (C) gives the detailed results for the variation in Stiffness derivative.

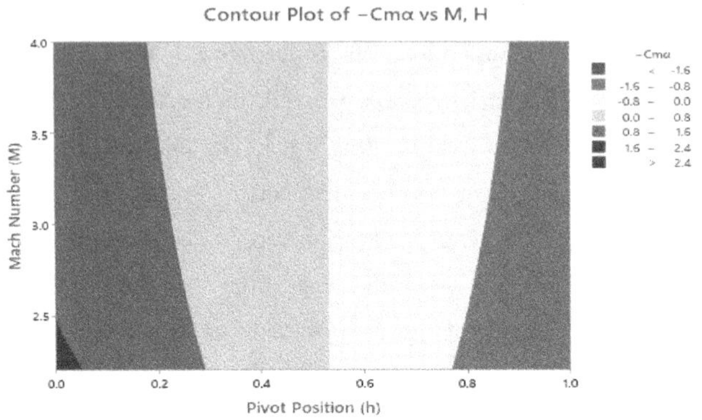

Fig. 5.1. 6 Contour plot for Stiffness Derivative with Mach number (M) and Pivot position (h)

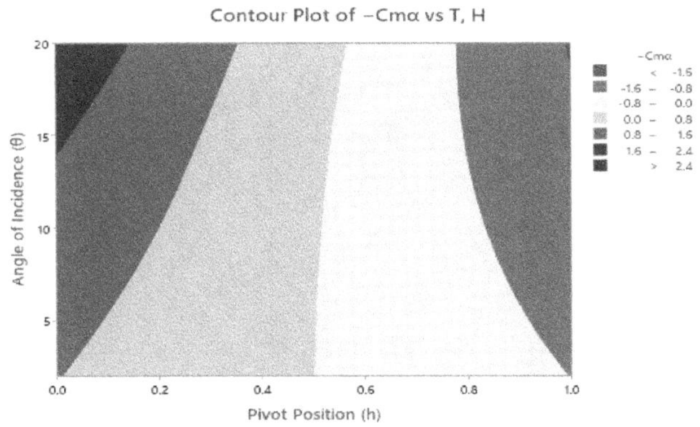

Fig. 5.1. 7 Contour plot for Stiffness Derivative with Angle of Incidence (θ) and Pivot position (h)

5.1.5 Regression Analysis

Minitab software creates the regression model, which evaluates stiffness derivatives according to Mach number (M), angle of incidence (θ), and pivot location (h). Eq. (5.1.9) explains the regression model for the derivative of stiffness.

$$-Cm_{\alpha} = \begin{pmatrix} 2.9547 - 1.2574\ M + 0.06775\ T - 5.7696\ H + 0.1805\ M*M + 0.000794\ T*T - 0.0000\ H*H \\ -0.00905\ M*T + 2.3349\ M*H - 0.11151\ T*H - 0.00457\ M*M*M + 0.000028\ T*T*T \\ -0.0000\ H*H*H + 0.001345\ M*M*T - 0.28904\ M*M*H - 0.000121\ M*T*T \\ +0.002445\ M*T*H + 0.00000\ M*H*H - 0.000803\ T*T*H + 0.000000\ T*H*H \end{pmatrix} \quad (5.1.9)$$

5.1.5.1 Model Summary

Table 5.3 presents the summary of the regression analysis model for Mach number (M), angle of incidence (θ), and pivot position (h). A statistical performance metric called adjusted R^2 is used to evaluate a regression model's goodness of fit. It shows the degree to which the independent variables may be able to explain the variability in the dependent variable. The regression model set forth in the study appears to explain the given data to the best extent, as evidenced by an R^2-square (adjusted) value of 99.99%.

Table 5. 3 Summary of Regression model.

S	R-square	R-square(adj)	R-square(pred)
0.0097679	99.99%	99.99%	99.99%

5.1.5.2 Literature-Based Validation of Current Work

As illustrated in Fig. 5.1.7(a), Fig.5.1.7(b), and Fig. 5.1.7(c), the findings of the current investigation have been verified by comparison with results found in the literature. Fig. 5.1.7(a), Fig.5.1.7(b), and Fig. 5.1.7(c) depict the findings of the analytical expression of stiffness derivative examined by Ghosh and Mistry [9] given by Eq. (5.1.7). The results for stiffness derivative obtained using the regression model given by Eq. (5.1.9). The obtained results indicate that the results of the stiffness derivative with analytical analysis and the current research work exhibit excellent agreement, as seen in Fig. 5.1.7(a), Fig.5.1.7(b), and Fig. 5.1.7(c).

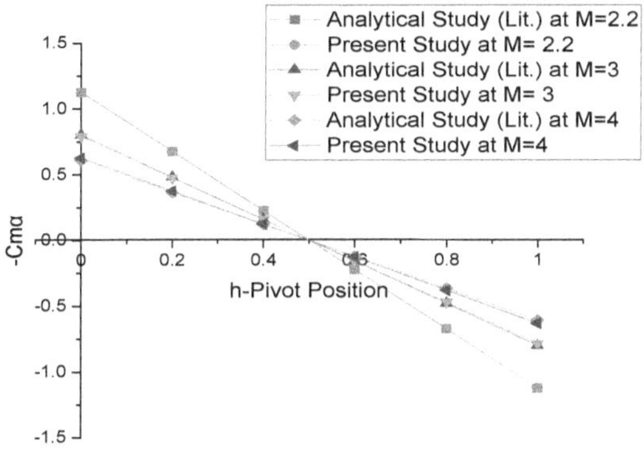

(a) at θ = 2°

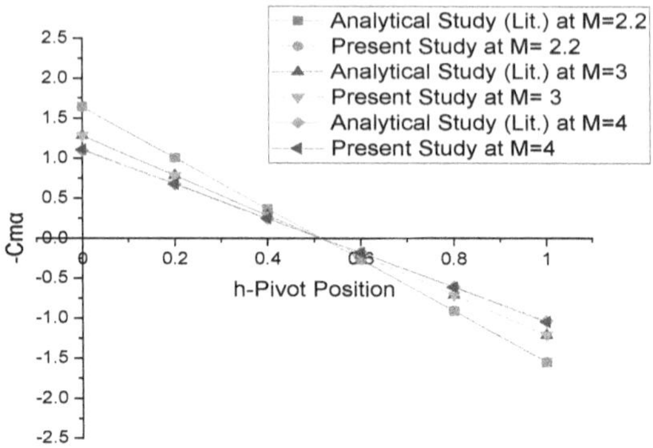

(b) at θ = 10°

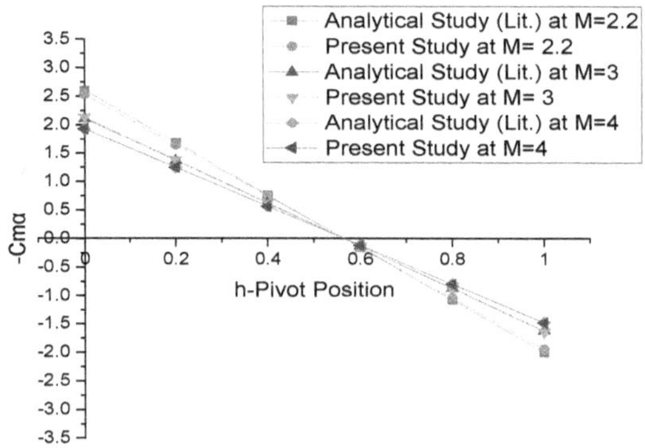

(c) at θ = 20°

Fig. 5.1. 8 Stiffness Derivative variation with Pivot Position (h)

5.1.6 Outcomes

Using the Ghosh piston theory, an analytical determination of the stiffness derivative is made, and the findings are contrasted with the current research findings. The results would be in close agreement with the present investigations, thus cleansed by analytical results. The research findings demonstrate diverse applications for angle of incidence (θ) and Mach number (M). Based on the data, the stiffness derivative falls linearly as the Mach number (M) grows at each pivot location (h). The stiffness derivative grows with the angle of incidence (θ) at each pivot location. Furthermore, it is found that the variables with the most significant effects on the change in stiffness derivative are the Mach number (M), incidence angle (θ), and pivot location (h). Owing to the significant cost associated with wind tunnel testing, these findings contribute to the development of aerial vehicles. These findings can help enhance the design of aerospace vehicles. The current investigation yields outstanding results with substantial computational ease. The theory only holds when the shock wave is attached. The boundaries of this theory can be stretched by adding the effects of viscosity and reflection of waves.

5.2 Evaluation of Damping Derivative of a 2D Wedge at Supersonic Speed

Analysis:

The geometry of the planar wedge illustrated in Fig. 5.1.1

The flat plate aerofoil should observe the small pitch oscillations near the pivot point, determined by its length (L) and wedge angle (θ). The variable x_0 denotes the distance from the apex. Eq. (5.2.1) gives the velocity of the piston at any instant, taking into account the distance x and the angle of attack α.

$$U_p = U_\infty \sin\alpha + q(x - x_0) \tag{5.2.1}$$

Eq. (5.2.2) gives the Mach number 'Mp' of the piston as,

$$M_p = M_\infty \sin\alpha + \frac{q(x - x_0)}{a_\infty} \tag{5.2.2}$$

Lighthill [18] examined the precise isentropic formula for piston pressure outlined in Eq. (5.2.3), a power series linked to velocity. The speed of the piston must be equal to or lower than the speed of sound in a free stream to keep the isentropic condition. The concept aligns with Lighthill's [18] Piston theory, which emphasizes the importance of minor disturbances.

$$\frac{P}{P_\infty} = 1 + \gamma\frac{U_p}{a_\infty} + \frac{\gamma(\gamma+1)}{4}\left(\frac{U_p}{a_\infty}\right)^2 + \frac{\gamma(\gamma+1)}{12}\left(\frac{U_p}{a_\infty}\right)^3 \tag{5.2.3}$$

5.2.1 Piston Theory

Lighthill [18] created the piston theory for oscillating air foils at supersonic and hypersonic velocities. According to Ghosh [25], the z-axis, where the velocity is least, can be seen as separate from the wedge strip aligned with the centreline. The integration of strip theory with Ghosh's [26] similitude method resulted in the creation of the "Piston analogy," specifically focusing on the piston Mach number "Mp." This leads to using Ghosh's piston theory instead of the strong shock expansion concept proposed by Lighthill [18] or Miles [23].

$$\frac{P}{P_\infty} = 1 + A(M_P)^2 + A M_P \sqrt{B + (M_P)^2} \qquad (5.2.4)$$

where $A = \frac{\gamma(\gamma+1)}{4}$, $B = \left(\frac{4}{\gamma+1}\right)^2$ and $P\infty$ and γ are the free stream pressure and specific heat ratio, respectively.

The strips are processed independently at various points along the span, with the wedge angle representing the wing. At this point, the piston Mach number and the flow deviation are allowed to increase. The incidence and wing angles are equivalent. The piston analogy has been widened to include supersonic flow past, as seen in Eq. (5.2.5). The related expression is below. With a uniform incidence angle and wing angles, the angle 'ø' here denotes the shock angle.

$$\frac{P}{P_\infty} = 1 + A\left(\frac{M_P}{\cos\phi}\right)^2 + A\left(\frac{M_P}{\cos\phi}\right)\sqrt{1 + B\left(\frac{M_P}{\cos\phi}\right)^2} \qquad (5.2.5)$$

5.2.2 Pitching Moment Derivative

When a wedge rotates about its axis and moves forward by a distance along the x direction, the angle of attack is referred to as α_0. The pressure and piston motion remain consistent on the wing side across a strip, with a length of 2L at point x. Eq. (5.2.5) describes the nose-down moment, assuming no pressure on the lee surface.

$$m = -\int_0^L P(x - x_0)\, dx \qquad (5.2.6)$$

The aerodynamic damping derivative is of the form as,

$$-Cm_q = \frac{1}{\frac{1}{2}\rho_\infty U_\infty L^3}\left(\frac{-\partial m}{\partial q}\right)_{\alpha=\alpha_0, q=0} \qquad (5.2.7)$$

Here ρ_∞ denotes the density of the free stream.

By equating Eq. (5.2.2) and Eq. (5.2.5) in Eq. (5.2.6) and applying DUIS in Eq. (5.2.6). Also defining, $x_0 = hL\cos^2\alpha_0, C = L\cos\alpha_0$ and $S_1^1 = \frac{M_\infty \sin\alpha_0}{\cos\phi}$ the desired relation of the damping derivative of a plane wedge is in the form of,

$$-Cm_q = \left[\frac{\gamma+1}{M_\infty}\right]\left[\frac{F(S_1^1)}{\cos\phi\cos^3\alpha_0}\right]\left[\frac{1}{3} - h\cos^2\alpha_0 + h^2\cos^4\alpha_0\right] \tag{5.2.8}$$

Where,

$$F(S_1^1) = 2S_1^1 + \frac{B + 2(S_1^1)^2}{\sqrt{B + (S_1^1)^2}} \tag{5.2.9}$$

5.2.3 Methodology of Design of Experiment

A structured approach using experiments, known as Design of Experiments (DOE), is employed to understand the impact of factors on a process and its outcomes. This method is crucial for creating, enhancing and refining processes. Below are the principles and stages involved in DOE:

i. Setting the objective which clearly states the problem.
ii. Identifying the factors to be tested with their specific values and desired range.
iii. Choose the appropriate type of experimental design according to the objectives.
iv. Use the specified design matrix to experiment, ensuring it follows the experimental plan.
v. Analyse the data using statistical approaches like regression and ANOVA to gain relevant insights and find significant components.
vi. Draw practical conclusions and recommendations based on the outcomes of the analysis. Provide graphical representations to help readers visualize the data and validate the outcomes.
vii. Interpret the results to understand the main and interaction effects and find the optimal conditions.
viii. Verify the results and validate the model against the actual results.

Apply the optimal conditions or changes to the process or product to implement the model.

The following statistical software packages are available to facilitate the design and analysis of DOE: Minitab, Statistica, SPSS, SAS, Design-Expert, Stat graphics, Prisma, R, and Action for Microsoft Excel. Mach Number (M), Wedge angle (θ), and Pivot Position (H) are the factors used by DOE. Creating the experiment requires taking into account the six different levels of pivot position (H) at level 10 while also factoring in Mach number (M) and incidence angle (θ). Table 5.4 depicts the data's factor, factor level, and values.

Table 5. 4 Full factorial: Level and the factors of the design of experiments

Factor	Type	Levels	Values
Mach Number (M)	Fixed	10	2.2, 2.4, 2.6, 2.8, 3.0, 3.2, 3.4, 3.6, 3.8, 4.0
Wedge Angle (T)	Fixed	10	2, 4, 6, 8, 10, 12, 14, 16, 18, 20
Pivot Position (H)	Fixed	6	0.0, 0.2, 0.4, 0.6, 0.8, 1.0

5.2.4 Results and Discussion

5.2.4.1 Major impact plots of Mach numbers for the Damping derivative $(-C_{m_q})$

In Fig. 5.2.1, the primary influence of the damping derivative on the wedge is illustrated. Fig. 5.2.1 presents the average damping derivative values across speed ranges from Mach 2.2 to 4.0. These measurements were taken at angles of incidence, ranging from 2° to 20°, and for pivot positions varying from h = 0.0 to 1.0. The findings indicate that the damping derivative values exhibit a specific trend as the Mach number increases. Additionally, there is a notable difference in the mean damping derivative values observed between Mach numbers 3.0 and 4.0 compared to those at Mach numbers 2.2 and 2.8. Overall, the damping derivative values gradually decline with the increase in Mach number.

5.2.4.2 Major impact plots of the angle of incidence for the Damping derivative $(-C_{m_q})$

Fig. 5.2.2 vividly showcases how changes in damping along the wedge impact the overall performance. This figure depicts the correlation between mean damping derivative values and the angle of incidence (θ) spanning from 2° to 20°. The Mach numbers featured in the diagram range from 2.2 to 4.0, while the pivot locations vary between 0.0 and 1.0. The findings reveal that the damping derivative tends to increase with the angle of incidence. Notably, this variation is more pronounced for incidence angles between 10° and 20° than those between 2° and 8°. In essence, the damping derivative progressively rises as the incidence angles increase.

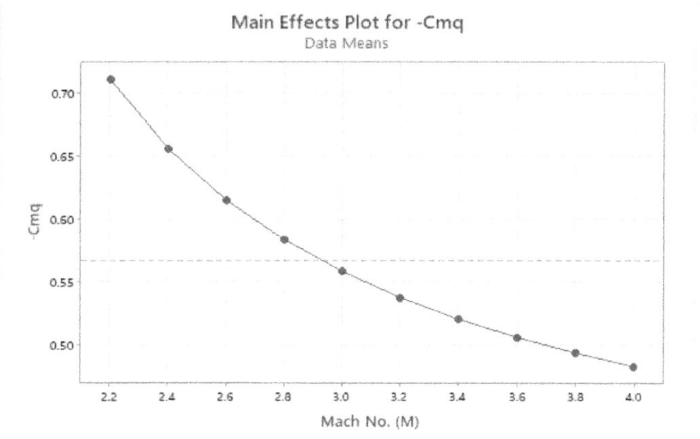

Fig. 5.2. 1 Major Effect of Mach numbers on the Damping Derivative for the wedge.

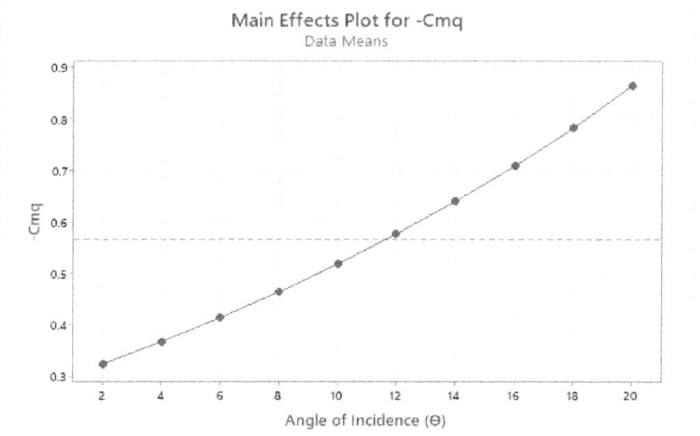

Fig. 5.2. 2 Major Effect of incidence angle on Damping Derivative for the wedge

5.2.4.3 Major impact plots of pivot position for the Damping derivative ($-C_{m_q}$)

Fig. 5.2.3 illustrates the influence of the damping derivative on the wedge. The graph showcases the average values of the damping derivative at pivot positions of 0.0, 0.2, 0.4, 0.6, 0.8, and 1.0. These values are derived from a range of Mach numbers (M) from 2.2 to 4.0 and incidence angles (θ) varying from 2° to 20°. The results indicate a decrease in the damping derivative when transitioning from the pivot point h = 0.0 to h = 0.6. In contrast, the damping derivative increases from the pivot position h = 0.6 to h = 1.0.

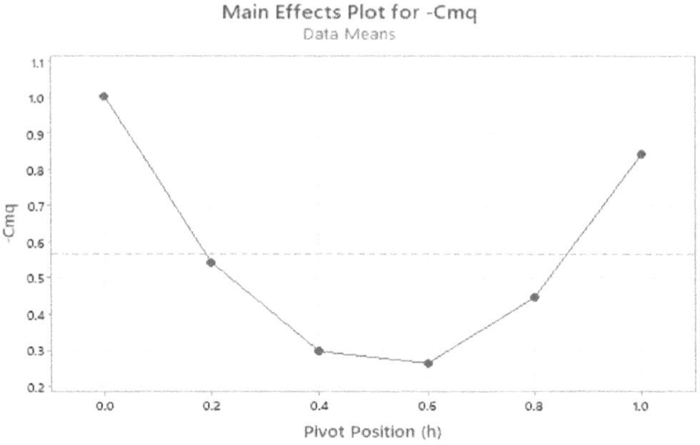

Fig. 5.2. 3 Major effect of the pivot position on the Damping Derivative over the wedge

5.2.4.4 Graph of Interaction among Incidence Angle (θ), Mach Number (M), and Pivot Position (h) for Damping derivative ($-C_{m_q}$)

Fig. 5.2.4 illustrates the interaction diagram of the damping derivative showcasing the relationship among Mach number (M), angle of incidence (θ), and pivot position (h). The data reveals that the damping derivative initially decreases, reaching its minimum value at the center of pressure before rising again from the pivot position h = 0.6. This trend persists across angles. Additionally, it is noted that the center of pressure falls within the range of h=0.4 to 0.6. Moreover, as the incidence angle increases, the damping derivative's magnitude also grows at pivot positions.

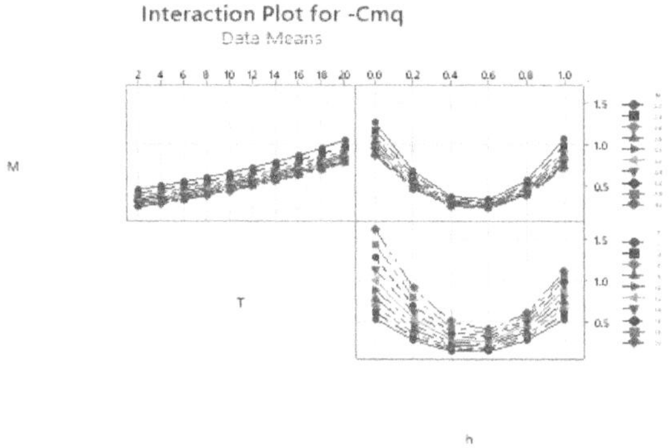

Fig. 5.2. 4 Interaction plot on Damping Derivative over the 2D wedge

5.2.4.5 The contour plot for the Damping derivative ($-C_{m_q}$) over wedge

Contour plots show the damping derivative over a wedge-shaped structure. In Fig. 5.2.5, it shows how the damping derivative varies with changes in (Mach Number) M and pivot position (h). It indicates that with increasing Mach number at positions, the magnitude of damping derivative reduces. The damping derivative shows an interesting trend attaining its minimum when h ranges from 0.4 to 0.6 where the center of pressure is located. The damping derivative reduces for every Mach number, after which it reaches its value within h values of 0.4 to 0.6 and eventually starts to rise again from the pivot position of h = 0.6 to 1. The contour plot in Fig. 5.2.6 shows the relationship between the damping derivative Mach number (M) and angle of incidence (θ). On a detailed note, from the plot, we can see that as the angle of incidence increases, the magnitude of the damping derivative also increases for several Mach numbers and locations. The damping derivative attains its value at higher Mach numbers at angles of incidence. Mach number and angle of incidence affect the damping derivative, but the angle of incidence appears to have a significant effect on its variations. The Appendix – (D) gives the detailed results for the variation in Damping derivative.

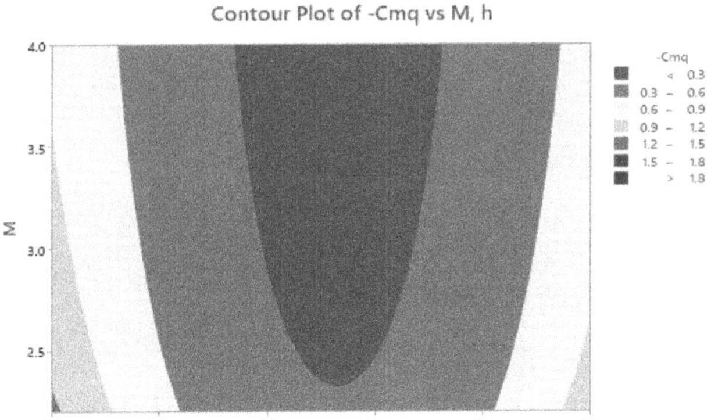

Fig. 5.2. 5 Contour plot for Damping Derivative with Mach number and Pivot position

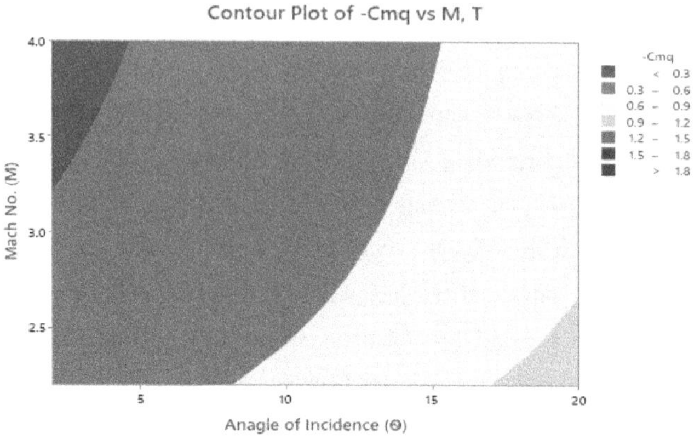

Fig. 5.2. 6 Contour plot for Damping Derivative with Mach number and Angle of Incidence

5.2.5 Model of Regression

The assessment of damping derivatives through a regression model is carried out using Minitab software, taking into account variables like position (h), angle of incidence (θ), and Mach number (M). Eq. (5.2.10) outlines the regression structure for calculating the damping derivative.

$$-Cm_q = \left(\begin{array}{l} 2.770 - 1.712M + 0.04151T - 3.499H + 0.4298M*M + 0.001415T*T \\ + 2.9058H*H - 0.00600M*T + 0.8995M*H - 0.12850T*H \\ - 0.03841\ M*M*M + 0.000012T*T*T + 0.0893\ H*H*H \\ + 0.000720\ M*M*T - 0.0583M*M*H - 0.000097M*T*T \\ + 0.002486M*T*H - 0.5593M*H*H - 0.001755T*T*H \\ + 0.12727T*H*H \end{array} \right) \quad (5.2.10)$$

5.2.5.1 Summary of Regression Model

Table 5.5 gives a description of the regression model that is estimated within the full dataset to determine its validity concerning pivot location (h), angle of incidence (θ), and Mach number (M). For this analysis, adjusted R^2, a statistical measure, is used to judge the capability of the regression model in representing damping derivatives at supersonic. This degree of adjusted R^2 defines how independent variables account for variation in dependent variables. With a value of 99.35 for R2, the research indicates that the regression model is well-suited for data involvement.

Table 5. 5 Summary of Regression model.

S	R-square	R-square(adj)	R-square(pred)
0.0279476	99.37%	99.35%	99.31%

5.2.5.2 Validation of Current Work with Literature

The results of the current study have been confirmed by comparison with data from previous studies, as shown in Figs. 5.2.7(a), 5.2.7(b), and 5.2.7(c). The results of the analytical expression of the damping derivative, as determined by Eq. (5.2.8) and studied by Ghosh and Mistry [25], are shown in Figs. 5.2.7(a), 5.2.7(b), and 5.2.7(c). Eq. (5.2.10) gives the results for damping derivative by using the regression model. Figs. 5.2.7(a), 5.2.7(b), and 5.2.7(c) show an excellent agreement for the results of the damping derivative with analytical analysis and the current research activity, as indicated by the obtained results.

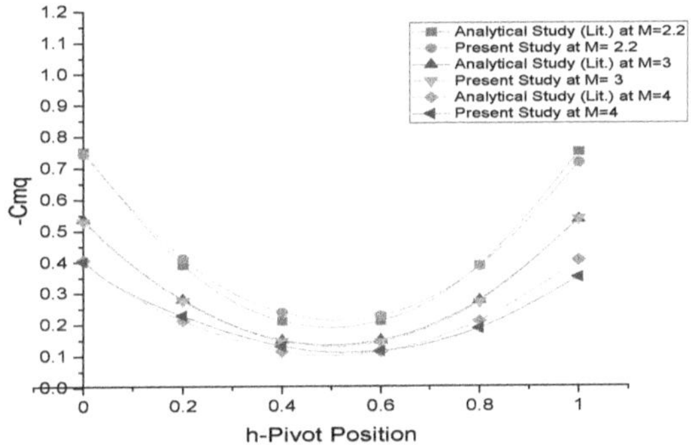

(a) at $\theta = 2°$

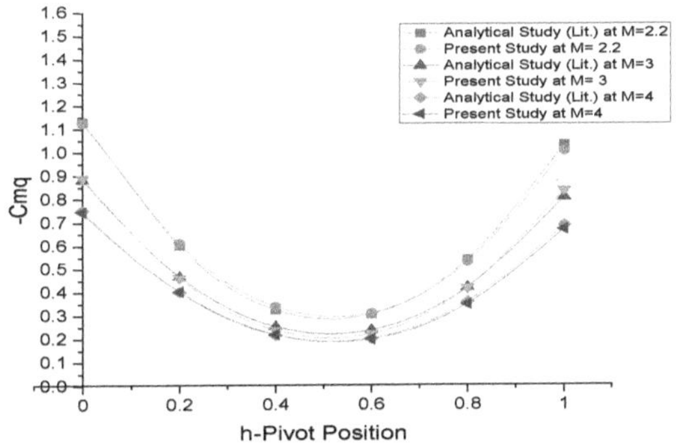

(b) at $\theta = 10°$

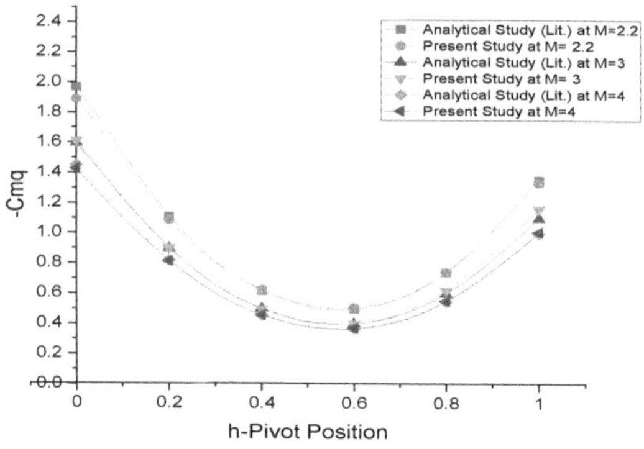

(c) at θ = 20°

Fig. 5.2. 7 Damping Derivative variation with Pivot Position (h)

5.2.6 Outcomes

Applying the Ghosh piston theory, an analytical calculation of the damping derivative is conducted, and its findings are contrasted with the conclusions of previous research. There is a good correlation between the current work's results and the analytical results. The findings of the research show various applications for angle of incidence (θ) and Mach number (M). Results showed that the damping derivative values changed considerably with semi-vertex angle and Mach number. The value for damping derivative changes at all pivot point positions with increasing Mach number. The magnitude of the damping derivative has found to increase with angle incidence. The result showed that at low Mach numbers and high angles of attack, the damping derivative had higher magnitude.

On the contrary, damping derivative exhibits lower value at minimum angle of attack and higher Mach Numbers. These insights play a role in advancing the progress of aerial vehicle design due to the high costs associated with wind tunnel testing. They hold the potential to enhance the design of aerospace crafts. The present study produces excellent outcomes with significant computational ease. This notion only holds if the shock wave is connected. This theory can consider the effects of viscosity and wave reflection to broaden its application and make it more accurate.

5.3 Evaluation of Stiffness Derivative of a 2D Wedge at Hypersonic Speed

Analysis:

The geometry of the planar wedge illustrated in Fig. 5.1.1

Concerning its length (L) and wedge angle (θ), the flat plate aerofoil should detect tiny pitch fluctuations close to the pivot point. The variable' x_0' represents the distance from the apex. Eq. (5.3.1) provides the piston's velocity at any given time,

$$U_P = U_\infty \sin\alpha + q(x - x_0) \qquad (5.3.1)$$

Eq. (5.3.2) gives the expression for piston number 'Mp' as,

$$M_P = M_\infty \sin\alpha + \frac{q(x - x_0)}{a_\infty} \qquad (5.3.2)$$

Lighthill [18] investigated the accurate isentropic equation for piston pressure expressed as a power series linked to velocity. To maintain the isentropic condition, the piston's speed must not exceed the speed of sound in a free stream. This idea is consistent with Lighthill's [18] piston hypothesis, which is based on the principle of minor disturbances.

$$\frac{P}{P_\infty} = 1 + \gamma\frac{U_p}{a_\infty} + \frac{\gamma(\gamma+1)}{4}\left(\frac{U_p}{a_\infty}\right)^2 + \frac{\gamma(\gamma+1)}{12}\left(\frac{U_p}{a_\infty}\right)^3 \qquad (5.3.3)$$

5.3.1 Piston Theory

The piston theory for oscillating airfoils at supersonic and hypersonic velocity was first developed by Lighthill [18]. At high angles of attack, Ghosh [24] discovered a similarity and two characteristics for a shock wave on the edge of vibrating delta wings. According to Ghosh [24], the wedge strip oriented in line with the centerline can be seen to be separate from the z-axis, where the velocity is lowest. Strip theory and Ghosh's [24] similarity method were combined to create the "Piston analogy," which focuses on the piston Mach number "Mp." Consequently, Ghosh's piston theory is applied instead of the powerful shock expansion concept proposed by Lighthill [18] or Miles [23]. Eq. (5.3.4) gives the pressure ratio as,

$$\frac{P}{P_\infty} = 1 + E(M_P)^2 + EM_P\sqrt{F + (M_P)^2} \qquad (5.3.4)$$

With a uniform incidence angle and wing angles, the angle 'ø' denotes the shock angle. Where:

$$E = \frac{\gamma(\gamma+1)}{4}, F = \left(\frac{4}{\gamma+1}\right)^2 \qquad (5.3.5)$$

and $P\infty$ and γ are the free stream pressure and specific heat ratio, respectively.

5.3.2 Pitching Moment Derivative

When a wedge rotates about its axis and moves forward by a distance along the x direction, the angle of attack is referred to as α_0. The pressure and piston motion remain consistent on the wing side across a strip, with a length of 2L at point x. Eq. (5.3.6) describes the nose-down moment, assuming no pressure on the lee surface.

$$m = -\int_0^L P(x - x_0)\, dx \qquad (5.3.6)$$

The corresponding expression of aerodynamic stiffness derivative as:

$$-Cm_\alpha = \frac{1}{\frac{1}{2}\rho_\infty U_\infty^2 L^2}\left(\frac{-\partial m}{\partial \alpha}\right)_{\alpha=\alpha_0, q=0} \qquad (5.3.7)$$

Here, the density and velocity of the sound for free stream as ρ_∞ and a_∞. By combining Eq. (5.3.2) and Eq. (5.3.4) in Eq. (5.3.6) and DUIS is being applied. Also, by setting:

$$x_0 = hL\cos^2\alpha_0, C = L\cos\alpha_0 \text{ and } S_1^1 = \frac{M_\infty \sin\alpha_0}{\cos\phi} \qquad (5.3.8)$$

The expression for required stiffness derivative of planar wedge in pitch will take the form as,

$$-Cm_\alpha = \left[\frac{\gamma+1}{M_\infty \cos\alpha_0 \cos\phi}\right] F(S_1^1) \left[\frac{1}{2} - h\cos^2\alpha_0\right] \qquad (5.3.9)$$

Here:

$$F(S_1^1) = 2S_1^1 + \frac{F + 2(S_1^1)^2}{\sqrt{F + (S_1^1)^2}} \qquad (5.3.10)$$

5.3.3 Design of Experiment Methodology

The Design of Experiment (DOE) systematically explores how different variables impact desired outcomes. It's a valuable method for gathering and analyzing data in scenarios where factors can be adjusted to assess their influence on results. By simultaneously investigating inputs, DOE can uncover interactions that may go unnoticed when examining one variable at a time. This technique finds applications across manufacturing, process optimization, product development, and engineering decision-making. Industries, including marketing, chromatography, energy, healthcare, pharmaceuticals, and food production, have adopted DOE. This methodology allows for exploring combinations of factors rather than focusing on a few selected ones. It applies to both real-world processes and computer simulation models. DOE is a tool for identifying input factors and understanding how they relate to output variables in various contexts. The approach plays a role in developing, improving, and fine-tuning procedures.

Outlined below are the key principles and steps in the Design of Experiments (DOE):

i. Defining a clear objective and outlining the issue.
ii. Identify the variables to be tested, specifying their values and desired ranges.
iii. Choosing a design framework like full factorial or fractional factorial to investigate combinations.
iv. Experiment with randomization, replication, and blocking to minimize bias and variability.
v. Gathering data in a manner that ensures precision and accuracy.
vi. Analyzing the data using methods such as ANOVA regression and factorial analysis for interpretation.
vii. Interpreting the results by examining effects and interactions and identifying optimal conditions.
viii. Validate the findings by comparing the model with actual outcomes.
ix. Implementing the identified conditions or adjustments in the process or product based on the model.

Various software tools can streamline the process of planning and assessing experiments, including Minitab, Statistica, SPSS, SAS, Design Expert, Stat Graphics, Prisma, R, and Action for Excel. In the context of the design of experiments (DOE), the primary considerations involve Mach Number (M), wedge angle (θ), and pivot position (H). When setting up the experiment, it's essential to evaluate six different levels of pivot position (H), Mach number (M) at level 11, and incidence angle (θ) at level 10. Details regarding the factor, factor level, and factor values are provided in Table 5.6.

Table 5. 6 Full factorial: Level and the factors of the design of experiments.

Factor	Type	Levels	Values
Mach Number (M)	Fixed	10	5.0, 5.5, 6.0, 6.5, 7.0, 7.5, 8.0, 8.5, 9.0, 9.5, 10.0
Wedge Angle (T)	Fixed	10	3,6,9,12,15,18,21,24,27,30.
Pivot Position (H)	Fixed	6	0.0, 0.2, 0.4, 0.6, 0.8, 1.0.

5.3.4 Results and Discussion

5.3.4.1 Major Effect graph of Mach number over Stiffness derivative ($-C_{m_\alpha}$)

The Fig. 5.3.1 presents the average stiffness derivative values and Mach numbers ranging from 5.0 to 10.0 with varying angles of incidence (θ) between 3° and 30° across all pivot locations h= 0.0 to 1.0. The findings indicate a decrease in the stiffness derivative values as there is a rise in the Mach number. Furthermore, a distinction can be observed in the mean stiffness derivative values for Mach numbers 8.0 to 10.0 compared to those for Mach numbers 5.0 and 7.5. Thus, the increasing Mach number shows a gradual decline in the stiffness derivative value.

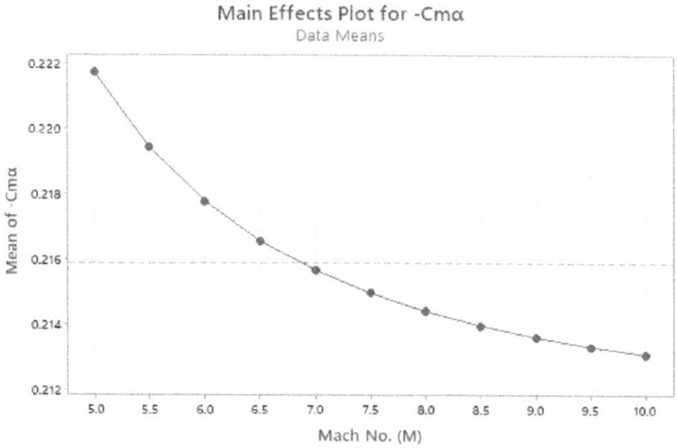

Fig. 5.3. 1 Effect of Mach number on the Stiffness Derivative across the 2D wedge

5.3.4.2 Major Effect graph of angle of incidence over Stiffness derivative ($-C_{m_\alpha}$)

The Fig. 5.3.2 displays the average stiffness derivative and angle of incidence (θ) values across a range of 3° to 30° for different Mach numbers (M) between 5.0 and 10.0 and pivot positions (h) varying from 0.0 to 1.0. The results indicate that the stiffness derivative tends to rise as the angle of incidence also rises. Notably, there is a more pronounced difference in the stiffness derivative for incidence angles between 18° and 30° than between 3° and 15°. Thus, the stiffness derivative exhibits an upward trend in value with higher incidence angles.

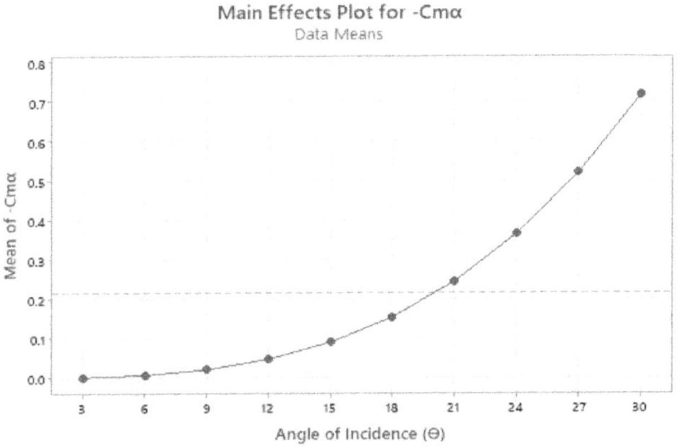

Fig. 5.3. 2 Effect of incidence angle on the Stiffness Derivative across the 2D wedge

5.3.4.3 Major Effect graph of pivot position over Stiffness derivative ($-C_{m_\alpha}$)

Fig. 5.3.3 illustrates the average stiffness derivative values at pivot points ranging from 0.0 to 1.0. The analysis considers varying Mach number (M) values of 5.0 to 10.0 and angles of incidence (θ) ranging from 3° to 30°. The results reveal a decrease in the stiffness derivative starting from the pivot point h = 0.0 and continuing to 1.0. Furthermore, the reduction in stiffness derivative is linear across this range.

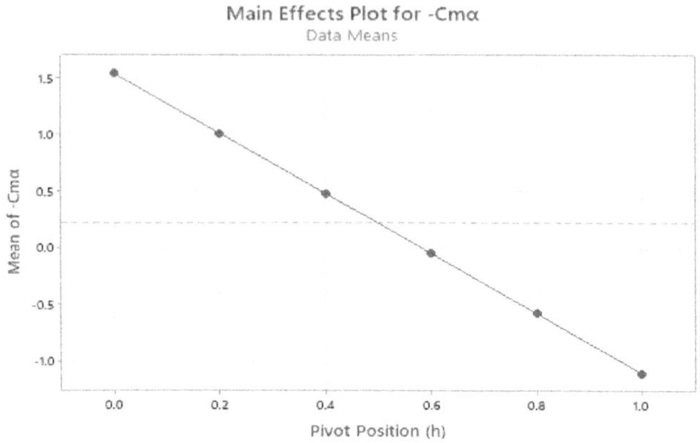

Fig. 5.3. 3 Effect of pivot position on the Stiffness Derivative across the 2D wedge.

5.3.4.4 Plot of Interaction among Incidence Angle (θ), Mach Number (M), and Pivot Position (h) for Stiffness derivative($-C_{m_\alpha}$)

The interaction plot amongst the parameters Mach number, angle of incidence, and pivot position is shown in Fig. 5.3.4. It is observed from this figure that the stiffness derivative generally shows a decreasing trend with increasing Mach number. The interaction between the Mach number and pivot position is strong, with stiffness derivative suffering significant reductions with increasing Mach number, particularly at higher pivot positions. The stiffness derivative is fairly stable over the range of Mach numbers considered.

Still, it shows more variability when interacting with changes in pivot position, suggesting a mild but noticeable impact on the overall aerodynamic stability. It is also observed that the pivot position exhibits the most significant influence on the stiffness derivative, showing substantial interaction effects with both Mach number and angle of incidence. As pivot position increases, the stiffness derivative generally decreases, with sharper declines observed when combined with high Mach numbers or varying incidence angles.

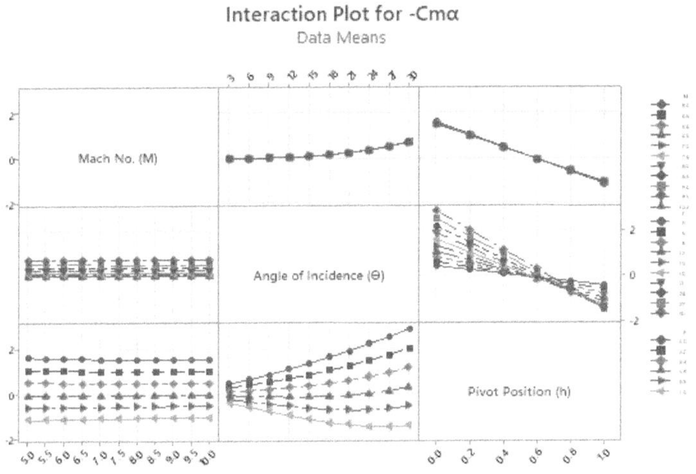

Fig. 5.3. 4 Plot of interaction for the Stiffness Derivative across the 2D wedge

5.3.4.5 The contour plot for the Stiffness Derivative ($-C_{m_\alpha}$) over wedge

The Fig. 5.3.5 highlights the contour plot showing the stiffness derivative concerning the Mach number (M) and angle of incidence (θ). The graph shows that the stiffness derivative of the Mach number remains relatively steady. However, there is a significant rise in its value as the angle of incidence increases, particularly at higher angles nearing 30°. According to this, the stiffness derivative is more responsive to changes in the angle of incidence rather than fluctuations in the Mach number within the examined ranges. Figure 5.3.6 displays the variation of stiffness derivative concerning Mach number (M) and pivot position (h). The stiffness derivative is generally unaffected by changes in Mach number, and shifts in the pivot location strongly impact it. The Appendix – (E) gives the detailed results for the variation in Stiffness derivative.

According to Fig. 5.3.6, the pivot position has a dominant influence on stiffness derivative compared to the Mach number within the analyzed range. The variability of the stiffness derivative concerning the pivot position (h) and angle of incidence (θ) is shown in Fig. 5.3.7. Fig. 5.3.7 depicts that as the pivot point and angle of incidence increase, the stiffness derivative is shown to climb considerably. This suggests that both parameters are highly relevant and that the effect becomes more pronounced when the pivot position is near 1.0 and the angle of incidence approaches 30°.

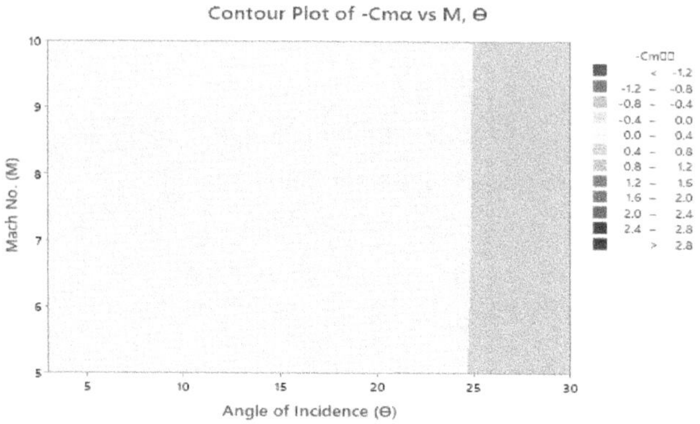

Fig. 5.3. 5 Contour plot for Stiffness Derivative with Mach number (M) and Angle of Incidence (θ)

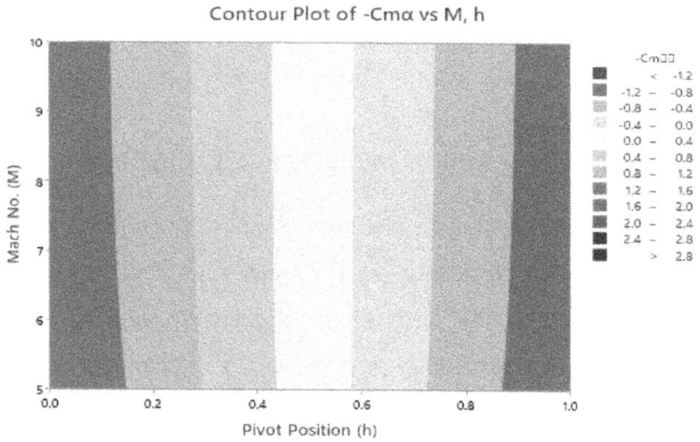

Fig. 5.3. 6 Contour plot for Stiffness Derivative with Mach Number (θ) and Pivot position (h)

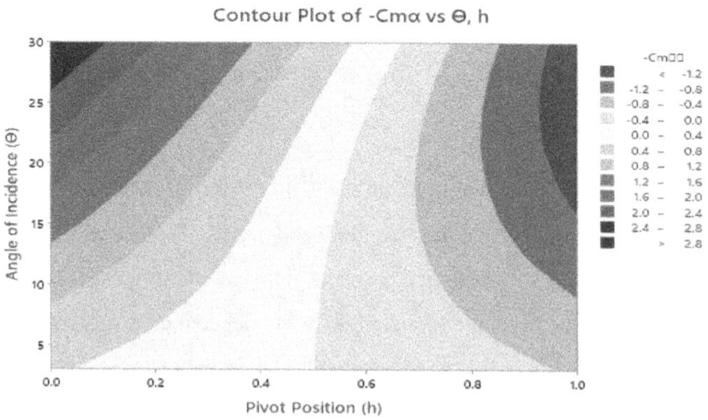

Fig. 5.3. 7 Contour plot for Stiffness Derivative with Angle of Incidence (θ) and Pivot position

5.3.5 Regression Analysis

A framework for evaluating the variation of stiffness derivatives concerning the Mach number (M), angle of attack (θ), and pivot location (h) is developed using the Minitab software. Eq. (5.3.11) describes the regression strategy to find these stiffness derivatives.

$$
-Cm_{\alpha} = \begin{pmatrix} 1.2366\ -0.2772\,M + 0.03208\ T - 2.4217\ h \\ +0.02717\,M*M + 0.001643\ T*T + 0.003142\ M*T \\ +0.5190\ M*h - 0.05570\ T*h - 0.001183\ M*M*M \\ +0.000058\ M*M*T - 0.04587\ M*M*h - 0.000064\ M*T*T \\ -0.009180\ M*T*h - 0.003191\ T*T*h + 0.000016\ M*M*M*M \\ +0.001507\ M*M*M*h + 0.000234\ M*M*T*h \\ +0.000115\ M*T*T*h + 0.000060\ T*T*T*h \end{pmatrix} \quad (5.3.11)
$$

5.3.5.1 Summary of Regression Model

Table 5.7 presents an overview of the regression analysis conducted for Mach number (M), angle of incidence (θ), and pivot position (h). To evaluate the effectiveness of the regression model in capturing the data, a statistical measure called adjusted R^2 is used. This measure reflects the extent to which the variables explain the variation in the dependent variable. In this research, the regression model is consistent with the data, showing an adjusted R^2 value of 98.35%, showcasing its precision.

Table 5. 7 Summary of Regression model.

S	R-square	R-square(adj)	R-square(pred)
0.0264776	98.27%	98.35%	98.31%

5.3.5.2 Validation of the Current Study Through Existing Literature

The validation of the results in this study is illustrated in Figures 5.3.8(a), 5.3.8(b), and 5.3.8(c) through a comparison with existing literature. The findings from the stiffness derivative analysis performed by Ghosh and Mistry [25] using Eq. (5.3.9) are displayed in these figures. Furthermore, the stiffness derivative results from the regression model outlined in Eq. (5.3.11) are also presented. These results show a close alignment between the stiffness derivative outcomes of the previous analysis and the current research, as depicted in Figures 5.3.8(a), 5.3.8(b), and 5.3.8(c).

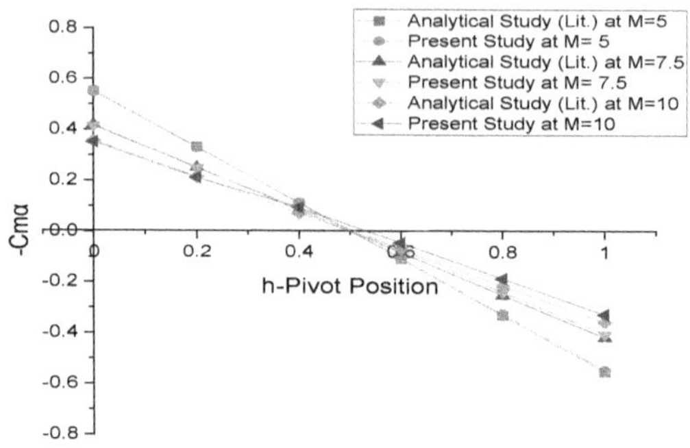

(a) at θ = 3°

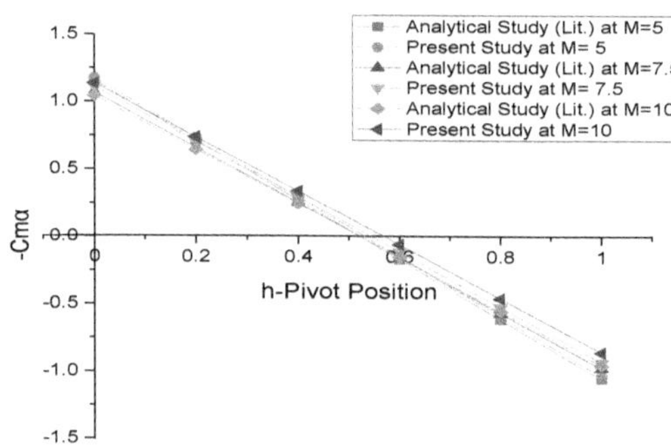

(b) at θ = 12°

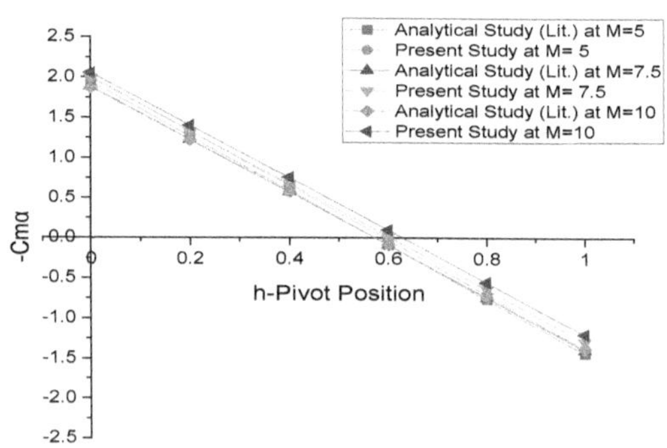

(c) at θ = 21°

Fig. 5.3. 8 Change in Stiffness Derivative with Pivot Position (h)

5.3.6 Outcomes

A detailed evaluation of the stiffness derivative is carried out through the Ghosh piston analogy, and the results are compared to those of previous research studies. The findings of this investigation show a significant correlation with the analytical outcomes. The study demonstrates applications across various angles of incidence (θ) and Mach numbers (M). The results indicate that the stiffness derivative diminishes with an increase in the Mach number (M) at all pivot positions (h). On the other hand, the stiffness derivative shows an increase with the angle of incidence (θ) at each pivot location. In addition, it has been observed that the factors that have the most significant influence on changes in stiffness derivative are the Mach number (M) incidence angle (θ) and pivot position (h).

Notably, the pivot position interacts strongly with the Mach number and incidence angle and significantly affects the variation of stiffness derivative. These findings are crucial for advancing vehicle development efforts, explicitly considering the high costs associated with wind tunnel testing. They hold promise for enhancing the design of aerospace vehicles. The current study yields results with remarkable computational efficiency. However, this theory only applies when the shock wave remains attached.

5.4 Evaluation of Damping Derivative of a 2D Wedge at Hypersonic Speed

Analysis:

The two-dimensional geometric representation of the planar wedge, illustrated in Fig. 5.1.1, is a straightforward visual aid for understanding the fundamental principles.

At the pivot position, the flat plate aerofoil should experience small pitch oscillations around the pivot point, defined by its length (L) and wedge angle (θ). The variable x_0 indicates the distance from the apex, and Eq. (5.4.1) reveals piston velocity at any instant by considering the distance x and incidence angle α.

$$U_P = U_\infty \sin\alpha + q(x - x_0) \quad (5.4.1)$$

Eq. (5.4.2) gives the Mach number 'Mp' of the piston as,

$$M_P = M_\infty \sin\alpha + \frac{q(x - x_0)}{a_\infty} \quad (5.4.2)$$

Lighthill [18] meticulously examined the precise isentropic equation about piston pressure, as interpreted in Eq. (5.4.3); this represents a power series correlated with velocity. The velocity of the piston must, however, be equal to or lower than the speed of sound in an unconfined stream to preserve the isentropic condition. Although this concept is congruent with Lighthill's Piston theory, it fundamentally emphasizes the importance of minor disturbances.

$$\frac{P}{P_\infty} = 1 + \gamma \frac{U_p}{a_\infty} + \frac{\gamma(\gamma+1)}{4}\left(\frac{U_p}{a_\infty}\right)^2 + \frac{\gamma(\gamma+1)}{12}\left(\frac{U_p}{a_\infty}\right)^3 \tag{5.4.3}$$

5.4.1 Piston Theory

Lighthill [18] initially formulated the piston theory to elucidate the dynamics of oscillating aerofoils at supersonic and hypersonic velocities; however, Ghosh [24] delineated notable similarities and two principal characteristics about shock waves at the peripheries of vibrating delta wings subjected to high angles of attack. According to Ghosh [24], the z-axis, where velocity remains minimal, can be considered distinct from the wedge strip aligned with the centerline. By synthesizing strip theory with Ghosh's [24] similitude method, the "Piston analogy" emerged, placing significant emphasis on the piston Mach number, designated as "Mp." This innovative approach catalyzed the preference for Ghosh's piston theory, superseding the robust shock expansion paradigms posited by Lighthill [18] or Miles [23]. Eq. (5.4.4) underscores that surface pressure (P) directly influences the inertia experienced by the wing's surface piston, thus highlighting the intricate interplay between aerodynamic forces and structural responses.

$$\frac{P}{P_\infty} = 1 + A(M_P)^2 + AM_P\sqrt{B + (M_P)^2} \tag{5.4.4}$$

where $A = \frac{\gamma(\gamma+1)}{4}$, $B = \left(\frac{4}{\gamma+1}\right)^2$

The strips are processed independently at various points along the span: the wedge angle represents the wing. At this juncture, the piston Mach number and the flow deviation are permitted to increase. The incidence and wing angles are equivalent; however, the piston analogy has been expanded to incorporate supersonic flow past, as illustrated in Eq. (5.4.5). The corresponding expression is provided below. With a uniform incidence angle and wing angles, the angle 'ø' here denotes the shock angle because this is crucial for understanding the dynamics involved.

$$\frac{P}{P_\infty} = 1 + A\left(\frac{M_P}{\cos\phi}\right)^2 + A\left(\frac{M_P}{\cos\phi}\right)\sqrt{1 + B\left(\frac{M_P}{\cos\phi}\right)^2} \quad (5.4.5)$$

5.4.2 Pitching Moment Derivative

When a wing rotates about its axis and progresses forward by a distance (along the x direction), the angle of attack is denoted as α_0. The pressure and piston motion on the wing side remain consistent across a strip with a length of 2L at point x. Eq. (5.4.6) elucidates the nose-down moment, although it assumes no pressure is exerted on the lee surface. This assumption is critical, however, because it simplifies the analysis.

$$m = -\int_0^L P\,(x - x_0)\,dx \quad (5.4.6)$$

The aerodynamic damping derivative is of the form as,

$$-Cm_q = \frac{1}{\frac{1}{2}\rho_\infty U_\infty L^3}\left(\frac{-\partial m}{\partial q}\right)_{\alpha=\alpha_0, q=0} \quad (5.4.7)$$

Here ρ_∞ denotes the density of the free stream.

By equating Eq. (5.4.2) and Eq. (5.4.5) in Eq. (5.4.6) and applying DUIS in Eq. (5.4.6). Also defining, $x_0 = hL\cos^2\alpha_0$, $C = L\cos\alpha_0$ and $S_1^1 = \frac{M_\infty \sin\alpha_0}{\cos\phi}$ the desired relation of the damping derivative of a plane wedge is in the form of,

$$-Cm_q = \left[\frac{\gamma+1}{M_\infty}\right]\left[\frac{F(S_1^1)}{\cos\phi\,\cos^3\alpha_0}\right]\left[\frac{1}{3} - h\cos^2\alpha_0 + h^2\cos^4\alpha_0\right] \quad (5.4.8)$$

Where,

$$F(S_1^1) = 2S_1^1 + \frac{B + 2(S_1^1)^2}{\sqrt{B + (S_1^1)^2}} \quad (5.4.9)$$

5.4.3 Design of Experiment (DOE) Methodology

The Design of Experiment (DOE) represents a structured and systematic methodology employed to elucidate the relationship between various factors that influence a process and the resultant output of that process. This methodology is commonly used in many applications such as engineering, manufacturing, and other types of research; it works to increase the efficiency of various processes and the quality of its end-point products. However, the complexities of the DOE methodology require careful consideration of a multitude of factors. Below is a brief guideline of the methodology.

i. **Define the Objective**: Clearly outline the problem and goals of the study.
ii. **Finding Factors and Levels**: Identifying the factors to be tested with their specific values and desired range.
iii. **Selection of Experimental Design**: Choosing a design framework like full factorial or fractional factorial to investigate combinations
iv. **Conduct the Experiment**: Apply randomization, replication, and blocking to minimize bias and variability.
v. **Collect Data**: Collection of accurate and precise data systematically.
vi. **Analyze the Data**: To interpret results, use of the tools like ANOVA, regression, and factorial analysis.
vii. **Interpret Results**: Understand the main and interaction effects
viii. **Validation of the Model**: Validate the findings by comparing the model with actual outcomes.
ix. **Implement Changes**: Apply the optimal conditions to refine the process or product.

This methodology provides a robust framework for decision-making and process improvement.

The following statistical software packages are available to facilitate the design and analysis of DOE: Minitab, Statistica, SPSS, SAS, Design-Expert, Stat graphics, Prisma, R, and Action for Microsoft Excel. The factors employed in DOE are Mach Number (M), Wedge angle (θ), and Pivot Position (h). The creation of the experiment necessitates consideration of the six distinct levels of pivot position (h) at level 10 while concurrently incorporating Mach number (M) and incidence angle (is same as wedge angle) (θ). Table 5.4.1 illustrates the factor, factor level, and factor values associated with the data. However, one must be acutely aware of the interplay between these variables because the outcomes are sensitive to slight input variations. Although these software tools simplify the process, this does not negate the complexity of the underlying statistical principles.

Table 5. 8 Full factorial: Level and the factors of the design of experiments.

Factor	Type	Levels	Values
Mach Number (M)	Fixed	10	5.0, 5.5, 6.0, 6.5, 7.0, 7.5, 8.0, 8.5, 9.0,9.5,10.
Wedge Angle (T)	Fixed	10	3, 6, 9, 12, 15, 18, 21, 24, 27, 30.
Pivot Position (H)	Fixed	6	0.0, 0.2, 0.4, 0.6, 0.8, 1.0.

5.4.4 Results and Discussion

5.4.4.1 Major Impact Plots of Mach Numbers for the Damping Derivative ($-C_{m_q}$)

In Fig. 5.4.1, the predominant impact of the damping derivative on the wedge is interpreted. Fig. 5.4.2 outlines the mean damping derivative values across various speed ranges, specifically from Mach 5.5 to 10.0. These empirical measurements were obtained at incidence angles varying from 3° to 30° and for pivot positions fluctuating from h = 0.0 to 1.0. These findings suggest that as the Mach number increases, the average values of the damping derivative decrease steadily. Additionally, a notable difference in the mean damping derivative values observed between Mach numbers 5.5 to 7.5 compared to those at Mach numbers 8.0 to 10.0. Overall, the damping derivative values gradually decline with the increase in Mach number. The plot indicates the impact of flow characteristics at high speeds, such as changes in pressure distribution and shock wave interactions, which likely contribute to the observed decrease in (-Cmq).

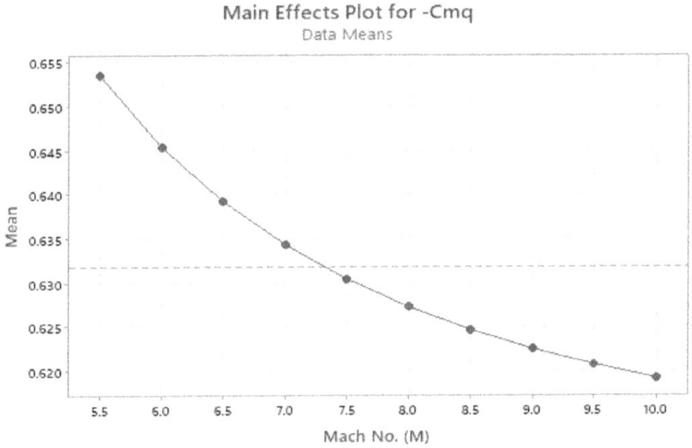

Fig. 5.4. 1 Displays the main effect of Mach numbers on the Damping Derivative over a 2D wedge

5.4.4.2 Major Impact Plots of the Angle of Incidence for the Damping Derivative ($-C_{m_q}$)

Fig. 5.4.2 demonstrates the effect of damping along the wedge on overall performance. It emphasizes the connection between average damping derivative values and the angle of incidence (θ) from 3° to 30°. The Mach numbers illustrated in the graph range from 5.5 to 10.0, and the pivot positions extend from 0.0 to 1.0. The findings suggest that the damping derivative typically rises as the angle of incidence increases. This rise is especially significant for angles from 18° to 30°, demonstrating a more pronounced increase than in the 3° to 15° range. Generally, the damping derivative gradually rises as the incidence angle increases. This data is critical for ensuring stability across various angles, especially for high-performance aircraft or missiles operating at extreme temperatures. The diagram shows a clear dependence of (-Cmq) on the angle of incidence, with more substantial damping effects observed at higher angles. This trend underscores the importance of considering angle-dependent damping in aerodynamic stability assessments and design optimizations.

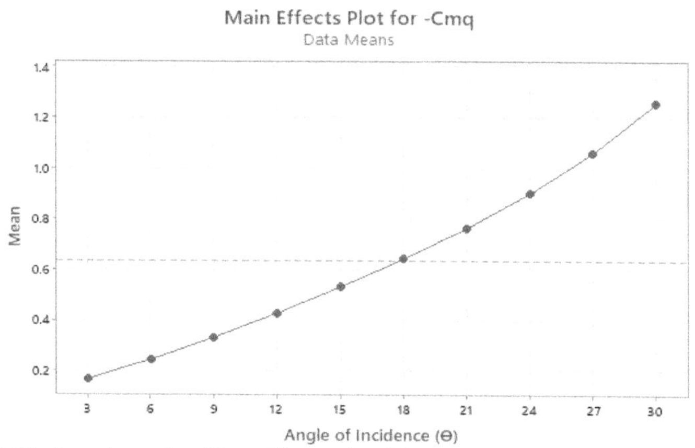

Fig. 5.4. 2 Displays the main effect of the angle of incidence (θ) on the Damping Derivative over a 2D wedge

5.4.4.3 Major Impact Plots of Pivot Position for the Damping Derivative ($-C_{m_q}$)

Fig. 5.4.3 illustrates the interaction diagram of the damping derivative as a function of three variables: Mach number (M), angle of incidence (θ), and pivot position (h). Considering collective effects of the aforementioned factors and their own interactivity on the damping derivative. In this respect, it is found that the damping derivative is mostly affected by two parameters: the pivot position and the angle of incidence; while the Mach number has less influence comparatively. The non-linear relationship of the damping derivative with respect to the pivot position suggests that aerodynamic stability is very much dependent on the very specific pivot position. This suggests a complex interplay of factors, although it is essential to recognize the pivotal role of the pivot position because it fundamentally determines stability outcomes.

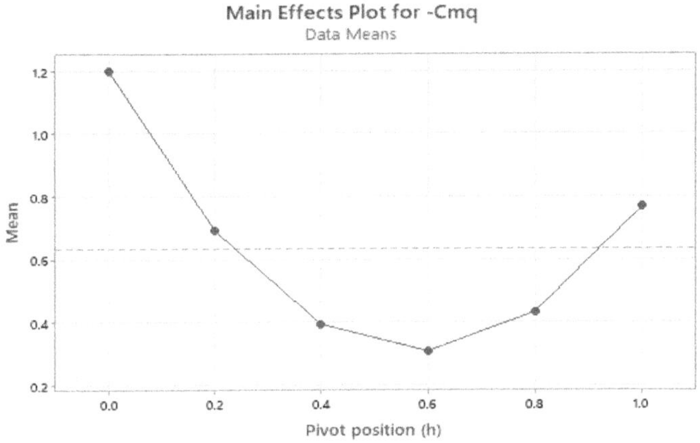

Fig. 5.4. 3 Displays the main effect of pivot position (h) on the Damping Derivative over a 2D wedge

5.4.4.4 Graph of Interaction among Incidence Angle (θ), Mach Number (M), and Pivot Position (h) for Damping Derivative ($-C_{m_q}$).

Fig. 5.4.4 illustrates the interaction diagram of the damping derivative as a function of three variables: Mach number (M), angle of incidence (θ), and pivot position (h). The plot highlights the combined effects of these factors on damping derivative and their interactions. The plot illustrates that the damping derivative is predominantly affected by the pivot position and the angle of incidence; however, the Mach number exerts a comparatively lesser influence. The non-linear relationship between the damping derivative and the pivot position indicates that aerodynamic stability significantly relies on the specific pivot position. This suggests a complex interplay of factors, although it is essential to recognize the pivotal role of the pivot position because it fundamentally determines stability outcomes.

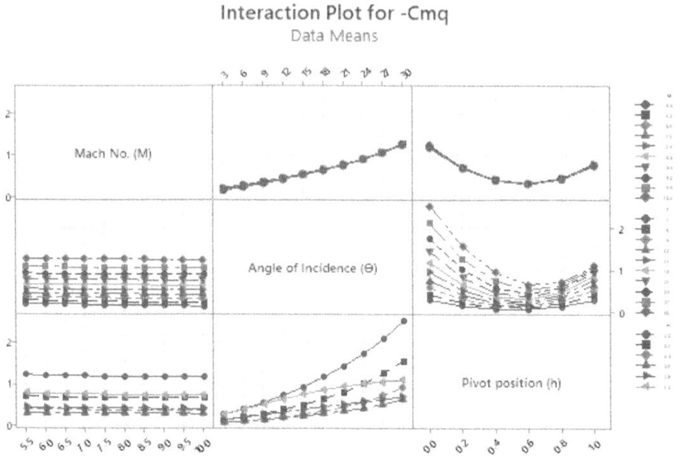

Fig. 5.4. 4 Displays the Interaction plot on Damping Derivative over the 2D wedge

5.4.4.5 The Contour Plot for the Damping Derivative ($-C_{m_q}$) over Wedge

In the Fig. 5.4.5 is the contour profile for Mach number (M) and angle of incidence (θ). One can see from this figure that the data indicate an increase in the damping derivative with increase in both Mach number (M) and angle of incidence (θ), but the effect of angle of incidence is predominant. This brings to focus the importance of angle of incidence (θ) regarding aerodynamic stability and damping characteristics. The contour plot for angle of incidence (θ) and pivot position (h) is given in Fig. 5.4.6. This damping derivative is seen to vary as a function of pivot position (h) and Angle of Incidence (θ). The plot illustrates that the damping derivative exhibits a non-linear relationship with both pivot position and angle of incidence. The lowest damping derivative occurs around h= 0.5 and for θ = 10° to 20°, while extreme pivot positions and higher angles of incidence amplify the damping effects. Fig. 5.4.7 displays the contour plot of Mach number (M) and pivot position (h). From Fig. 5.4.7, shows that the damping derivative is predominantly influenced by pivot position (h) with minimum sensitivity to Mach number (M). The lowest values occur around h=0.4 to h=0.6, while higher values are observed at the extreme pivot positions. This contour plot highlights the critical role of aerodynamic stability and damping coefficient characteristics. The Appendix – (F) gives the detailed results for the variation in Stiffness derivative.

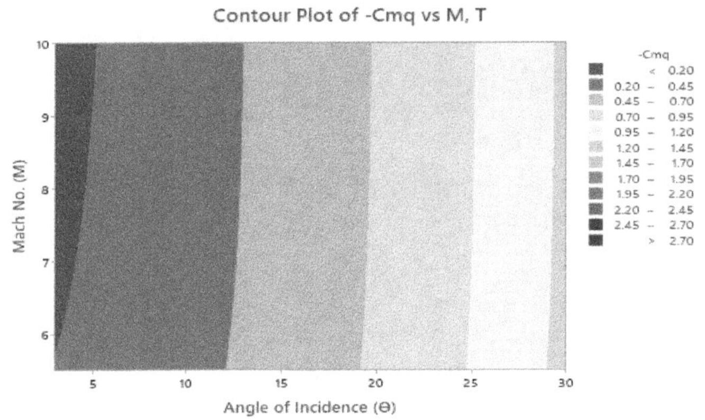

Fig. 5.4. 5 Contour plot for Damping Derivative with Mach number and Angle of Incidence Pivot position

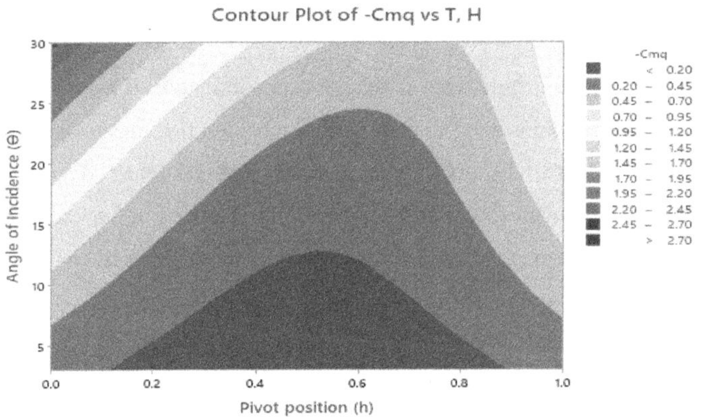

Fig. 5.4. 6 Contour plot for Damping Derivative with Mach number and Angle of Incidence Pivot position

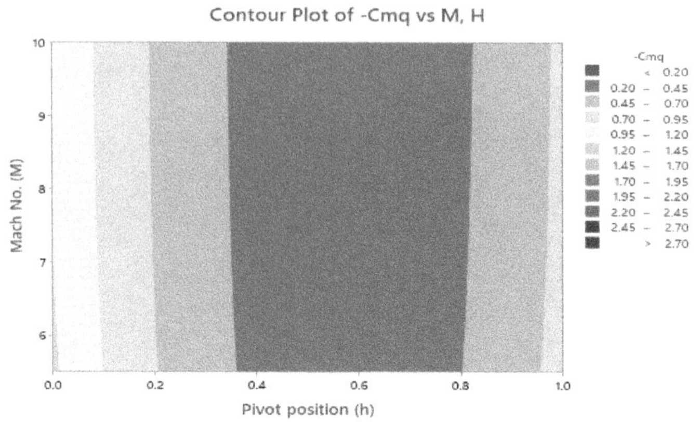

Fig. 5.4. 7 Contour plot for Damping Derivative with Mach number and Angle of Incidence Pivot position

5.4.5 Model of Regression

The evaluation of damping derivatives using a regression model is executed by employing Minitab software, which considers variables such as position (h), angle of incidence (θ), and Mach number (M). Eq. (5.4.10) outlines the regression framework to compute the damping derivative. The model provides a structured approach; the accuracy of the results may fluctuate because of various influencing factors. This elaborate relationship underscores the necessity for meticulous analysis and interpretation of the data.

$$-Cm_q = \begin{pmatrix} 0.619 - 0.1045\ M + 0.01864\ T - 0.8860H + 0.00861M * M + 0.001272T * T + 0.7292\ H * H \\ +0.001466\ M * T + 0.0419\ M * H - 0.09561\ T * H - 0.000271\ M * M * M + 0.000019\ T * T * T \\ -0.000029\ M * M * T - 0.000631\ M * M * H - 0.000025\ M * T * T + 0.000219\ M * T * H \\ -0.03310\ M * H * H - 0.002632\ T * T * H + 0.130833\ T * H * H \end{pmatrix} \quad (5.4.10)$$

5.4.5.1 Model Summary

Table 5.9 presents a comprehensive overview of the regression analysis model designed to scrutinize the intricate relationship between pivot location (h), angle of incidence (θ), and Mach number (M). In this study, a critical statistical measure adjusted R^2 is employed to evaluate the efficacy of the regression model in accurately representing damping derivatives at hypersonic speeds. This metric elucidates the extent to which the independent variables can explain variability in the dependent variable. With an R^2 value of 97.45%, the current research indicates that the regression model effectively encapsulates the underlying data.

Table 5. 9 Summary of Regression model.

S	R-square	R-square(adj)	R-square(pred)
0.0329476	97.37%	97.45%	97.41%

5.4.5.2 Validation of Current Work with Literature

The findings of the current study have been validated through a comparative analysis with data obtained from previous studies, as depicted in Fig. 5.4.8(a), Fig. 5.4.8(b), and Fig. 5.4.8(c). The analytical expression regarding the damping derivative, as expressed in Eq. (5.4.8) and examined by Ghosh and Mistry [25], is displayed in Fig. 5.4.8(a), Fig. 5.4.8(b), and Fig. 5.4.8(c). Furthermore, the damping derivative outcomes achieved by employing the regression model described in Eq. (5.4.10) are likewise shown in these figures. Fig. 5.4.8(a), Fig. 5.4.8(b), and Fig. 5.4.8(c) demonstrate a significant correspondence in the findings of the damping derivative with both the analytical analysis and the current research investigation.

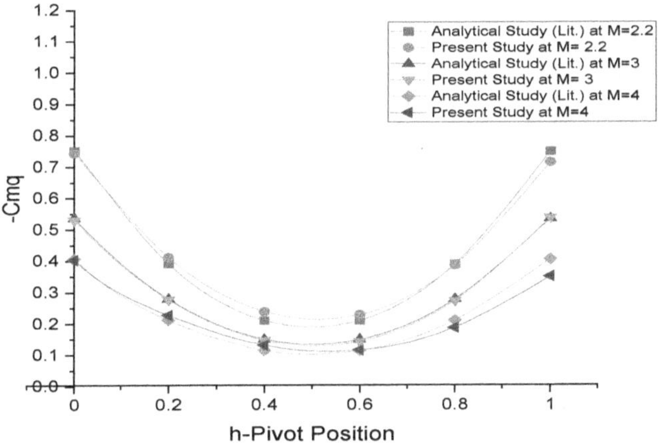

(a) at θ = 3°

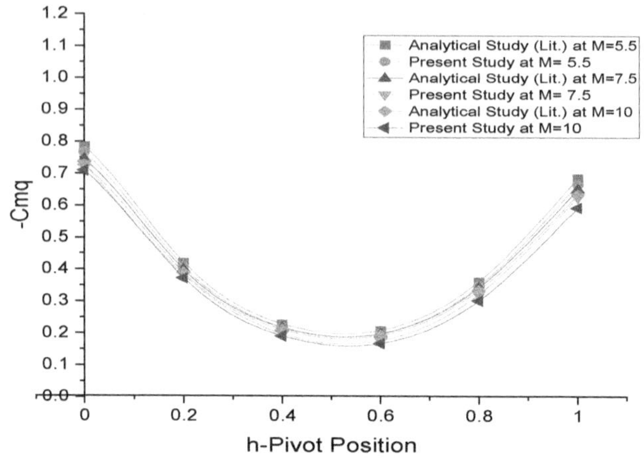

(b) at θ = 12°

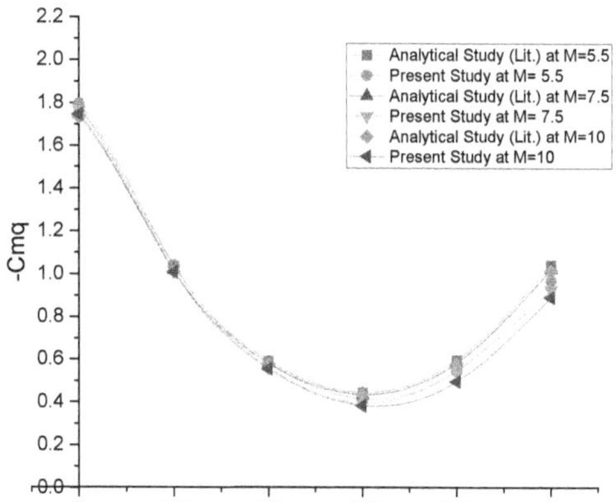

(c) at θ = 24°

Fig. 5.4. 8 Damping Derivative ($-Cm_q$) variation with Pivot Position (h)

5.4.6 Outcomes

Following Ghosh's piston theory, an analytical computation of the damping derivative has been undertaken; its findings are compared against the conclusions drawn from earlier investigations. The current study has an impressive degree of synergy with the analytical insights. Findings from the research indicate various applications where angle of incidence (θ) and Mach number (M) are variables. The results show that variations in damping derivative are highly dependent on both the Mach number (M) and the angle of incidence (θ). At each pivot position, the damping derivative value decreases as the Mach number increases. The extent of the damping derivative increases with the incidence angle. The results reveal that damping derivatives have an impressive value with high angles of attack and low Mach numbers.

On the other hand, the damping derivative exhibits a lower value for higher Mach numbers and lower angles of attack. The current research yields marvelous results with considerable computational simplicity. This concept is valid solely if the shock wave is linked.

www.ingramcontent.com/pod-product-compliance
Lightning Source LLC
LaVergne TN
LVHW010739061126
851857LV00014B/1550